Why Are We Producing Biofuels?

WHY ARE WE PRODUCING

BIOFUELS

Shifting to the Ultimate Source of Energy

ROBERT C. BROWN

AND TRISTAN R. BROWN

Brownia LLC
AMES, IOWA

Dr. Robert C. Brown is Anson Marston Distinguished Professor of Engineering and holds the Gary and Donna Hoover Chair in Mechanical Engineering at Iowa State University (ISU). He is internationally recognized for his research on the conversion of biomass into energy, fuels, and chemicals. Tristan R. Brown is an attorney who is currently employed as a research associate at Iowa State University's Bioeconomy Institute. He has published in both the fields of economics and engineering on topics ranging from greenhouse gas emissions policy to global trade law to engineering economics.

Brownia LLC, Ames 50014

Printed in the United States of America

18 17 16 15 14 13 12   1 2 3 4 5

ISBN-13: 978-0-9840906-3-1
ISBN-10: 0-9840906-3-0

Library of Congress Control Number: 2012904552

Illustrations by Trevor Brown and Carolyn Brown
Book design by Trevor Brown

www.brownia.com

You become responsible, forever, for what you have tamed.

— Antoine de Saint-Exupéry
Le Petit Prince, 1943

Contents

Acknowledgements

In the fall of 2008, one of the authors (RCB) was invited by Gregory Geoffroy, President of Iowa State University (ISU) at that time, to give a lecture on biofuels as part of the ISU Presidential Lecture Series. President Geoffroy recognized as early as 2002 the important role that biofuels could play in the nation's energy future and he invested considerable personal effort in raising the visibility of "biorenewables" at both ISU and across the state of Iowa, which was soon to become the leading biofuels producer in the United States. By 2008 ISU was making major contributions to the field through research, workforce development, and outreach. The intent of the lecture that fall was to explain biofuels, warts and all, to the university community and the interested public. Because the rapid growth of the biofuels industry was not without prominent critics, RCB entitled his talk "Why are We Producing Biofuels?" Immediately after the lecture, Professor Lawrence Johnson, who was one of the framers of the biorenewables program at ISU (and who RCB credits with coining the term "biorenewables"), suggested the lecture be expanded into a book. With encouragement from his family, RCB began to work on the book that winter break, anticipating it would be completed in a year. Before long, TRB joined the project and all but two chapters were drafted by early 2010, just a little behind schedule. But another book project, a new graduate course, and a couple of large research initiatives conspired to bring the project to a near standstill. Meanwhile, the field was rapidly evolving, threatening obsolescence of the material already prepared. Late in 2011, we decided to make a concerted push to finish the book in early 2012, which included extensive updating of the chapters already written and drafting the remaining chapters.

The authors would like to acknowledge several people who helped make this book possible. Most prominently, Trevor Brown, with assistance from Carolyn Brown, prepared the illustrations that appear in this book along with the cover art. He also designed the layout of the book and personally typeset it. Finally, Trevor provided all the services associated with privately publishing the book. Bruce Dale, Lawrence Johnson, Jill Euken, and Brian Gravelin read early manuscripts and provided valuable suggestions for improving its content, for which I am grateful. Of course, the authors take responsibility for any errors or omissions found in the book.

RCB, who during the time he was writing the book was also the director of the Bioeconomy Institute and the Center for Sustainable Environmental Technologies at ISU, would never have attempted this project nor seen it through completion without the knowledge that he had many talented people backing him up: Becky Staedtler, Jill Euken, Ryan Smith, Marge Rover, Patrick Johnston, Lysle Whitmer, Jordan Funkhouser, Jill Cornelis, Diane Meyer, Chris Knight, Jan Meyer, Diane Love, and an army of graduate and undergraduate students who worked in the laboratories. He also sends his love to wife Carolyn, who was always supportive of the goal of bringing the book to its conclusion.

Soon after this project commenced, TRB joined ISU's

Bioeconomy Institute as a research associate and instructor of a graduate course on biorenewables law and policy at ISU, which frequently suggested subject matter for the book. He gratefully acknowledges his frequent lunchtime conversations with Jerome Dumortier, Keith Evans, and Abhishek Somani, whose willingness to provide their professional opinions on subjects such as food versus fuel, indirect land-use change, and energy economics enhanced this book's discussion of those topics. He also wishes to thank his wife, Kate, for cheerfully enduring the numerous weekends and vacations that were spent working on this book.

Robert C. Brown
Tristan R. Brown
AMES, IOWA, February 2012

Abbreviations

ASP	Aquatic Species Program
BEV	battery electric vehicles
CARB	California Air Resources Board
CBP	consolidated bioprocessing
CC	combined cycle
CCS	carbon capture and sequestration
CFC	chlorofluorocarbons
CHP	combined heat and power
CNG	compressed natural gas
CTL	coal-to-liquids
DDGS	distillers' dried grains and soluble
DLUC	direct land use change
DME	dimethyl ether
E-10	ethanol (10% volume blend with gasoline)
E-85	ethanol (85% volume blend with gasoline)
ECVS	embedded carbon valuation system
EFA	essential fatty acids
EIA	(U.S.) Energy Information Agency
EISA	Energy Independence and Security Act
EPA	Environmental Protection Agency
EtOH	ethanol
FAME	fatty acid methyl ester
FCEV	fuel cell electric vehicle
FFV	flexible fuel vehicle

GDP	Gross Domestic Product
GH2	compressed hydrogen gas
GHG	greenhouse gases
GPY	gallons per year
HDI	Human Development Index
HEC	herbaceous energy crop
HEV	hybrid electric vehicle
HFCS	high fructose corn syrup
HTP	hydrothermal processing
ICE	internal combustion engine
IGCC	integrated gasification/combined cycle
ILUC	Iindirect land use change
kWh	kilowatt-hour
LCFS	Low Carbon Fuel Standard
LH2	liquefied hydrogen
LNG	liquefied natural gas
LPG	liquefied petroleum gas
LWR	light water reactor
MSW	municipal solid waste
MTBE	methyl tertiary butyl ether
NAAQS	National Ambient Air Quality Standards
NAFTA	North American Free Trade Agreement
NASA	National Aeronautics and Space Administration
NBB	National Biodiesel Board
NG	natural gas
NREL	National Renewable Energy Laboratory
OPEC	Organization of Petroleum Exporting Countries
PEM	proton exchange membrane
PHEV	plug-in hybrid electric vehicle
PV	photovoltaics
QE	quantitative easing
R^2	regression coefficient
RESS	rechargeable energy storage system

RFG	reformulated gasoline
RFS	Renewable Fuel Standard (2005 legislation)
RFS2	Renewable Fuel Standard (2007 legislation)
SNG	synthetic natural gas
SRWC	short rotation woody crop
SSCF	simultaneous saccharification and co-fermentation
SSF	simultaneous saccharification and fermentation
U.S. DOE	United States Department of Energy
UHC	unburned hydrocarbons
USDA	United States Department of Agriculture
VOC	volatile organic compounds
WTW	well-to-wheels

1

Answer in brief

Why are we producing biofuels? The answer, quite simply, is that we have few other options for achieving a sustainable energy future. Whereas product quality and cost were the primary metrics of the past, future fuels will have to meet additional metrics including environmental, social, and political sustainability. As the impact of fossil fuels on the health of humans and ecosystems become better understood and appreciated, national legislation and international agreements will constrain the emissions of nitrogen, sulfur, mercury and carbon dioxide from power plants, fuel refineries, factories, homes, and vehicles alike. In developed countries, as resource intensive industries move overseas and trade imbalances increase from the importation of petroleum, domestic labor markets will decline and communities suffer. Countries like the United States with dwindling domestic sources of petroleum will increasingly depend upon countries with poorly managed or politically manipulated energy supplies.

Transportation fuels, in the broadest sense, are forms of energy that are suitable for mobile power. Traditionally, these have been energy-rich liquids, like gasoline or ethanol, although gases and electricity can also serve as transportation fuels. Biofuels are transportation fuels produced from biomass, which is the generic term for any kind of plant material exploited as an energy source. Although corn ethanol and soy diesel often come to mind, biofuels can take many forms including fuels that are essentially indistinguishable from petroleum-derived gasoline or diesel.

Biofuels are not the only alternative to imported petroleum. Fuels can also be produced from other fossil fuels including coal, natural gas, tar sands, and oil shale. Except for biofuels, none of these alternatives has prospects for long-term sustainability as evaluated in terms of production costs, greenhouse gas emissions, water demand, impact on local communities, or infrastructure investment. Other kinds of renewable energy can also be converted into fuels. Although energy from wind and sunlight, geothermal heat, and water impounded behind dams is typically converted into electricity, it can be stored in batteries, electrolyzed to hydrogen, or used to drive chemical reactions of water and carbon dioxide that produce synthetic gasoline. Most of these renewable energy alternatives are more costly and less infrastructure compatible than biofuels. Conversion of biomass into biofuels is the best option for reducing use of petroleum and other fossil fuels.

The U.S. biofuels experience has been a complicated story of extremes, embodying both exuberant over-optimism and hyperbolic criticism. Ethanol and vegetable oil were widely accepted when internal combustion engines were first introduced in the late nineteenth century, but the following century saw declining interest as gasoline and diesel's lower costs and superior fuel qualities became apparent. Biofuels remained sidelined until the opening decade of the twenty-first century when it was discovered that petroleum-based MTBE (methyl tertiary butyl ether), added to gasoline to reduce smog-generating engine emissions, was contaminating groundwater. Ethanol was quickly adopted as an environmentally friendly replacement to MTBE by fuel producers eager to avoid costly litigation for potential environmental harm resulting from MTBE use. Congress soon recognized ethanol's ability to also serve as a homegrown gasoline substitute, a trait that meshed nicely with a growing national desire to decrease America's dependence on foreign petroleum. Legislation was soon passed mandating ethanol consumption on a national scale (the Renewable Fuel Standard, or RFS). The first decade of the twenty-first century witnessed an unprecedented boom in the U.S. biofuels industry with fuel ethanol production increasing by a factor of ten despite difficulties in integrating ethanol into the gasoline-based fuel infrastructure because of differences in fuel properties.

The corn ethanol boom in the United States was only the beginning of a national effort to substitute domestically produced biofuels for petroleum-based fuels. Recognizing that even the whole U.S. corn crop converted to ethanol could only replace about 15% of national gasoline consumption, agronomist have been developing alternative crops for biofuels production for over thirty years. More recently, there has emerged a growing resistance to the use of crops that were part of our food production system. Alternative biomass feedstocks include trees and tall prairie grasses grown in plantation stands as well as the residues from traditional crop production. Rather than producing crops rich in sugar, starch, protein, or oil, energy crops and crop residues primarily consist of the equivalent of dietary fiber. Known as cellulosic biomass, this material actually contains a mixture of complex carbohydrates protected by a glue-like chemical known as lignin. Although more challenging than turning cornstarch into ethanol, cellulosic biomass can be converted into so-called advanced biofuels, including "cellulosic ethanol," produced from sugars that are biologically extracted from cellulose, and "drop-in biofuels" that mimic the chemical profiles of gasoline and diesel. Encouraged by federal mandates for the production of advanced biofuels, venture capitalists, corporations and governments have invested billions of dollars in biofuels startup companies with business models built around cellulosic ethanol and drop-in biofuels. These investments have begun reaching the commercialization stage, with several advanced biofuels companies completing successful IPOs in 2010 and 2011.

The U.S. biofuels industry has largely been a legislative creation, fostered in its early years by substantial production subsidies and stiff tariffs on foreign biofuels and supported more recently by addition of a biofuels production mandate. The widespread and sustained bipartisan support of these initiatives by Congress indicates many legislatures believe that biofuels play an important part in meeting national policy goals. America's historical experience with petroleum shortages during times of military and economic conflict places energy security foremost among these goals, particularly as U.S. petroleum reserves have been steadily declining since the early 1970s. America's growing dependence on foreign petroleum has become particularly worrisome as a number of major petroleum-producing countries have over the last decade experienced internal instability or become hostile to the U.S., placing petroleum imported from those sources at risk of disruption. While recent technological developments have permitted the exploitation of unconventional petroleum reserves located either domestically or in countries with close trading partnerships with the U.S., the greater effort required to extract these reserves comes with increased costs, both economic and environmental.

The environmental impacts of petroleum extraction,

refining, and utilization are often cited as justification for policies that favor biofuels over fossil fuels. Petroleum-based transportation fuels are a leading source of U.S. anthropogenic greenhouse gas (GHG) emissions. This will soon be true for China and India as their populations become increasingly reliant upon automobiles. The atmospheric concentration of GHGs will continue to rise to historically unprecedented levels unless the nations of the world work in concert to reduce emissions of these gases, with uncertain consequences to the climate, the biosphere, and the world economy. Biofuels can dramatically reduce GHG emissions relative to gasoline and even cause net GHG sequestration if properly implemented. All biofuels are not created equal in this regard, with some controversial analyses ascribing greater GHG emissions to grain ethanol and soy biodiesel than gasoline and diesel fuel.

Petroleum continues to supply us with relatively inexpensive transportation fuels. Although cheap energy is often a pillar of economic expansion, this is not necessarily the case when the energy is imported petroleum. A growing body of economic data supports the theory that over reliance on imported petroleum was a leading cause of the 2008 Financial Crisis and subsequent Great Recession. Petroleum imports have historically been one of the largest contributors to the U.S. trade deficit, which has ballooned in the twenty-first century. This trade deficit and shrinking U.S. exports left foreign petroleum suppliers (and other major trading partners such as China) with few opportunities to spend their dollars other than to purchase U.S. government bonds. This growing demand for U.S. debt helped keep U.S. interest rates at historically low levels, prompting the growth of a bubble in the real estate market and the subsequent financial crisis that occurred when this bubble collapsed. Recognizing the role of trade deficits in precipitating economic crises, U.S. politicians have declared reducing the trade deficit to be an important goal over the next few years. Not withstanding these declarations, the trade deficit has continued to increase in the aftermath of the Great Recession, largely because petroleum imports account for most of this deficit. Efforts to prevent similar financial crises in the U.S. through a rebalancing of global trade will fail so long as imported petroleum contributes so much to the nation's trade deficit.

Over a decade ago President George W. Bush declared that it was time for America to end its addiction to imported petroleum. Although this sentiment is still widely shared by policy makers, the lack of a comprehensive national energy policy in the United States almost 40 years after the first oil shock of 1973 reflects the difficulties of selecting and developing alternatives. The process is complicated by the fact that we need to decide upon both primary energy sources and energy carriers for our transportation systems. Primary energy sources are forms of energy as they are found in nature, whether petroleum buried in geological deposits or wind circulating in the atmosphere. Rarely can these be directly used in internal combustion engines without further processing into cleaner, more convenient energy carriers such as gasoline or ethanol.

A surprising large number of primary energy sources could be substituted for imported petroleum to produce transportation fuels including other fossil fuels, nuclear energy, and renewable energy. Some of these resources, such as coal, gas hydrates and uranium are in reasonably good supply, although an often leery public will have to be convinced that they can be extracted and used without harm to the environment or their neighborhoods. This task has become even more daunting in the face of energy-related disasters such as the Fukushima nuclear reactor meltdowns in Japan. U.S. renewable energy reserves, which far outstrip U.S. reserves of fossil fuels that have accumulated over eons, have emerged as attractive alternative primary energy sources. The difficulty with renewable energy is not its scarcity but its diffuse nature compared to the fossil fuels that have powered our economy over the last century. The key to harnessing renewable energy is sufficiently concentrating it to power automobiles and aircraft.

Identifying appropriate primary energy sources to replace petroleum is only half the job, as this raw energy must be converted into an energy carrier compatible with our transportation systems. After all, we do not want to "fill up" our automobiles with loads of coal or biomass, although it has been done in the past. We settled upon gasoline and diesel for our internal combustion engines because these mixtures of hydrocarbons were the most logical energy carrier to produce from petroleum. Other energy carriers have been suggested for biofuels including: a substance coveted as an intoxicating beverage; a nitrogen-rich gas that farmers inject into the ground as fertilizer; a mixture of synthetic hydrocarbons formerly used to power the Nazi war machine; and an energy carrier that is not easily contained within a fuel tank but nevertheless is one of the leading contenders to replace petroleum-based fuels. Each of these energy carriers has their advocates, although only a few can be said to be ideal replacements for gasoline and diesel fuel.

In the United States only ethanol and biodiesel have been introduced as alternative energy carriers into the transportation system to any significant extent. Their emergence is the result of American agriculture's ability to produce corn and soya (soybeans) in excess of demand by either domestic or international markets. Corn grain is rich in starch, which is easily converted into sugars that can be fermented to ethanol. Soya is rich in vegetable oils (lipids), which is easily converted into a less viscous and more stable product known as biodiesel. Despite periodic run-ups in the price of gasoline and diesel that panic consumers and politicians alike, the United States has been blessed by historically low fuel prices compared to many other parts of the world, which has made it difficult for either corn ethanol or biodiesel to compete with petroleum-based fuels except in the presence of mandates and subsidies.

Brazil's biofuels experience has been quite different, and provides a prospectus on both the challenges and opportunities facing countries that adopt biofuels. Brazil built its biofuels industry on sugar cane, which is even easier to convert into ethanol than cornstarch. Launched in the 1970's as part of a national goal to achieve energy independence, the Brazilian government provided financial support to produce and use ethanol. This required mechanizing sugar cane production, constructing ethanol fermentation plants, and adopting a national flex-fuel infrastructure. These investments have enabled Brazil to become entirely self-reliant in its supply of transportation fuels, with ethanol comprising nearly half of the country's fuel supply and domestic petroleum supplying the rest. In contrast, the U.S. did not begin to produce ethanol fuel in earnest until the first decade of the twenty-first century and ethanol currently satisfies only 10% of gasoline demand. Biodiesel from soya is also part of Brazil's energy policy, but it plays only a small role compared to ethanol, as is also the case in the United States.

Cellulosic biomass is widely viewed as the "feedstock of the future" for the production of biofuels, promising a larger and less expense resource base than conventional crops. But cellulosic biomass is not only more difficult to convert into fuels than sugars, starch, and lipids, it contains considerably less energy for a given volume of material than traditional crops. Although this may appear a small disadvantage, it has major ramifications to the practical implementation of advanced biofuels. Modern agro-industry is premised upon high-density crops, which are more easily harvested, stored, and processed than "fluffy" materials like hay, cornstalks, or wood chips. Despite efforts to increase the density of cellulosic biomass by forming it into bales, for example, cellulosic biomass requires more manual labor and vehicles to move a given tonnage of biomass compared to traditional crops. Low density also means cavernous storage facilities must be constructed for year-round storage of biomass, which is usually harvested only once a year.

This rapid expansion has not come without a price. The advent of large-scale U.S. ethanol production was closely followed by a large rise in commodity crop prices in 2008. Unflattering headlines soon appeared in the press. The Economist, The New York Times, and TIME respectively wrote articles titled "Biofools", "Biofuel or Biofraud?" and "The Clean Energy Scam." Biofuels production was (and continues to be) blamed for rainforest destruction, starvation in the developing world and, in a wounding irony, global climate change. A United Nations official responsible for investigating human rights violations even went so far as to label biofuels a "crime against humanity" in comments widely carried by the international press. If the first decade of the twenty-first century was a headlong rush to biofuels, the second decade was a frenzied backlash to demonize them.

A more productive use of the present decade is to chart a course to sustainable energy supply that incorporates lessons learned in commercializing first generation biofuels. What did we learn? We demonstrated that it is possible to produce biofuels at commercial scale – fourteen billion gallons of ethanol per year from U.S. agriculture is, after all, a lot of fuel. We discovered that increasing domestic fuel production, even though only displacing 10% of gasoline supply, could shake up the energy industry, with gasoline refining facing a long decline in the United States and oil producing nations realizing that they are not the only players in fuel markets. In wondering how we will replace the other 126 billion gallons of gasoline consumed in the U.S, we recognize our profligacy in energy consumption. We also recognize that the first generation of biofuels is not the last generation of biofuels. This in no way denigrates the existing ethanol and biodiesel industries, whose leaders understand that innovation is critical to the future success of their technology-driven enterprises. Biofuels need to be improved in terms of infrastructure compatibility; optimal use of land to supply both food and fuel security; increasing the energy efficiency of biomass agriculture and biofuels production and utilization in vehicles; and achieving prices that are competitive with other fuels available to consumers. Three major pathways to advanced biofuels are generally recognized: lipid-based biofuels; cellulosic biofuels through biochemical processing; and cellulosic biofuels through thermochemical processing.

Lipids are attractive for production of biofuels because they can be easily upgraded into substitutes for gasoline, diesel, and aviation fuel using petroleum-refining technology. Offsetting this advantage, traditional sources of lipids, such as soya, are widely acknowledged as not yielding enough lipids per acre of land to be cost effective or land efficient. For this reason lipid-based fuels were for many years discounted for production of advanced biofuels. These prospects have recently improved as a result of challenges facing two important niche markets. Recent legislation in the European Union requires commercial aircraft flying into member nations to meet stringent greenhouse gas emission standards, which will be difficult to achieve except through the use of biofuels. The U.S. military, which is responsible for protecting U.S. energy supplies during times of war, has determined that its own reliance on imported petroleum compromises its strategic goals. The U.S. Navy, in particular, is undertaking plans to secure domestic sources of biofuels for ships, aircraft, and military vehicles in the next few years. For both aviation and military applications, only fuels that can meet stringent performance specifications can be considered. Lipid-based fuels, although expensive, can meet these standards and, by using existing refining infrastructure, can be put into production more quickly than other kinds of advanced biofuels.

These applications have increased interest in alternative sources of cheap and plentiful lipids. Microalgae, in particular, promises lipid yields that are several times greater than traditional oil-seed crops. Research into microalgae biofuels began in the 1970s as part of the U.S. Department of Energy's (DOE) "aquatic species" program. After twenty years the program was terminated in favor of expanding the development of cellulosic ethanol, which demonstrated more promising economics. Costs remain a major challenge for microalgae biofuels. Other lipid-rich feedstocks are also being developed. Plants that grow on wasteland or salt marshes are especially favored to avoid charges that these oils, some of which are edible, contribute to the "food vs. fuel" controversy that surround first generation biofuels.

Cellulose, the primary constituent of wood, grasses, and crop residues, closely resembles starch, which allows it to be broken down by enzymes into simple sugars and fermented by microorganisms to yield fuel molecules in a manner similar to the production of grain ethanol. However, the enzymes have a much tougher job of penetrating plant material to reach the strands of cellulose, which is more difficult to decompose than corn starch. Efficient fermentation also requires microorganisms that can use both glucose, a six carbon sugar, as well as a variety of five-carbon sugars released from hemicelluloses, another major form of carbohydrate in plant materials. Although cellulosic ethanol was the focus of most advanced biofuels research over the past 35 years, other biochemical pathways are now being explored that produce lipids, terpenes, and even hydrocarbons. These are attractive for the manufacture of "drop-in" biofuels, which are fully compatible with existing automotive and aviation transportation infrastructure.

Cellulosic biomass can also be converted to fuels by thermochemical pathways, which are distinguished from biochemical pathways in the use of heat and catalysts instead of enzymes and microorganisms to process biomass. They closely resemble today's power plants and petroleum refineries, with reactors operating at high temperatures and pressures and feedstocks requiring only minutes to be processed compared to many hours for typical fermentation processes. This resemblance is one of the chief attractions of thermochemical processing, suggesting close integration with existing petroleum refineries, which are increasingly underutilized in the United States. Although a number of thermochemical pathways have been devised, they can be roughly categorized as gasification, pyrolysis, or solvolysis, each of which yields intermediate products that can be upgraded to finished fuel or chemical products. Until recently, thermochemical technologies did not receive the same level of research support by the U.S. DOE. In part, this inattention reflected a common view that the biochemical platform would benefit from the revolution in biotechnology while the thermochemical platform relies on mature technology that offers few new advances. Cellulosic ethanol, in particular, holds promise for retrofitting grain ethanol plants to advanced biofuels, making it the natural successor to the first generation of biofuels plants. The U.S. DOE has moved to a more balanced portfolio of biochemical and thermochemical technologies for several reasons. Thermochemical technologies have historically focused on production of hydrocarbon products, which aligns well with the recent focus on drop-in biofuels. Recent analyses of the costs of advanced biofuels gives a decided edge in the near term to certain thermochemical pathways, which is important in jump starting the commercial production of advanced biofuels.

This book explores the opportunity to advance a sustainable energy future through the development of advanced biofuels. By examining the emergence of first generation biofuels and the kinds of technologies being developed for advanced biofuels, the book also articulates the challenges that must be overcome: Will the industry be driven by technological innovation or government policy? If not gasoline and diesel, what fuel will propel our vehicles? How is it that we are using food crops to produce motor fuels? What do the recent criticisms about biofuels portend for its future? How is it possible that a renewable fuel can contribute to global climate change? What kinds of biomass occur in sufficient quantity to help displace imported petroleum? How can these feedstocks be transformed into transportation fuels? What is the most likely future of fuels?

Although we may argue over the details of achieving a sustainable energy future, one thing seems certain. We must learn how to harness solar energy to produce both food and fuel in a manner that benefits all people in the coming century.

2

What are the origins of the biofuels era?

Although biofuels have a history dating back to Rudolph Diesel's development of the compression ignition internal combustion engine and Henry Ford's Model T automobile, it was not until 2005 that the biofuels era was officially launched in the United States. On August 8th of that year President George W. Bush signed into law the Energy Policy Act that set in motion a vast expansion of the renewable fuels industry. Long time biofuels boosters from the agricultural Midwest were joined by environmentalists, biotechnology companies, venture capitalists, consumers, university researchers, politicians, energy companies, and car manufacturers to map out what some observers described as the clean tech era. The promise of biofuels to reduce reliance on imported petroleum and help tackle the problem of global climate change was among the top news stories during the next two years.

The impetus was California's ban of methyl tertiary butyl ether (MTBE) as a fuel additive in January 2004, which was in response to growing concerns that defective fuel storage tanks were allowing this toxic, petroleum-derived chemical to contaminate drinking water supplies. Although other states had already started banning MTBE, California accounted for approximately one-third of the MTBE fuel additive market in the United States, giving it a major say in the future direction of national energy policy.

Ironically, it was concerns about air emissions from automobiles that led to the widespread blending of MTBE with gasoline as a pollution-control measure.[1] Although MTBE had been used at relatively low concentrations in gasoline as an octane enhancer since the late 1970s, its widespread adoption began in 1992 when the Environmental Protection Agency (EPA), responding to the Clean Air Act Amendment of 1990, implemented a "winter oxygenate fuels program" in regions of the U.S. that were not meeting National Ambient Air Quality Standards (NAAQS) for carbon monoxide (CO). Air emission monitoring in major cities found that CO pollution was particularly acute during winter months when temperature inversions trapped pollutants near the ground and cold engines did not operate as efficiently as during warmer weather. Engine research at that time demonstrated that fuels containing oxygen reduced CO and unburned hydrocarbons (UHC) emissions while boosting the octane rating of the fuel.

The Clean Air Act Amendment of 1990 addressed both CO and ozone pollution from automobiles.[1] Although the CO problem could be addressed with fuel oxygenates alone, ozone required a special blend of fuel components known as "reformulated gasoline" (RFG). Generally, these so-called "non-attainment regions" of the country were only in non-attainment for either CO or ozone but not both. The exception was the Los Angeles basin, where both CO and ozone were air pollution problems. Accordingly, the Clean Air Act Amendment of 1990 specified that RFG contain at least 2% oxygen by weight to help reduce CO emissions. The legislation did not specify the nature of the fuel oxygenate for manufacturing reformulated gasoline although many farm state interests hoped it would be ethanol. Reformulated gasoline requires only 7.5% by volume of ethanol to satisfy the fuel oxygenate requirement, which is easily met by E10, the standard blend of 10% by volume of ethanol with gasoline. However, these hopes were dashed when many states, especially near the coasts where petroleum refineries are concentrated, chose MTBE for their fuel oxygenate. Although 15% by volume of MTBE is required in reformulated gasoline to meet the fuel oxygenate requirement, MTBE was cheaper to ship and easier to blend than ethanol. MTBE dominated oxygenate production in the early years of the fuel oxygenate and reformulated gasoline programs.

Leaking fuel storage tanks changed all that. The ethanol industry widely anticipated the pending demise of MTBE, assuming that it would become the major provider of fuel oxygenate to meet the reformulated gasoline mandate. However, California and other states, backed by the petroleum industry and environmental groups, argued that advances in computer controlled, fuel injected engines had made fuel oxygenates obsolete in RFG.[2] Rather than substitute ethanol for MTBE in reformulated gasoline, these groups wanted to simply repeal the oxygenate requirement for RFG in the next federal energy bill. Since ethanol was more expensive than gasoline to produce, the loss of the RFG mandate would have brought an end to the fledgling grain ethanol industry. Nearly 13% of the nation's corn crop was being used to manufacture ethanol, an important outlet for grain markets. Farm state legislators strongly opposed any move that would weaken demand for corn. Legislators from states where petroleum refining and gasoline marketing were an important part of the economy were concerned that petroleum companies would face increased liability exposure if they continued to manufacture MTBE after the repeal of the reformulated fuel mandate. The petroleum industry sought legislative protection from lawsuits as they phased out MTBE manufacture over several years on the grounds that legislative mandate had forced them to use fuel oxygenates in gasoline in the first place. Legislators from states struggling to remediate MTBE contaminated water supplies were unwilling to make this concession, pointing out that legislation had not specifically prescribed the use of MTBE as the fuel oxygenate in reformulated gasoline.

The Energy Policy Act finally passed in 2005 after contentious debate. It replaced the reformulated fuel mandate with a renewable fuel standard.[2] In doing so, it implicitly acknowledged the limited environmental benefit of fuel oxygenates in RFG, whether MTBE or ethanol. Modern engine technology, by reducing the need for fuel oxygenates in gasoline, had trumped well meaning environmental policy that ultimately was judged to have caused more harm than good to the environment. The renewable fuel standard (RFS) that replaced the RFG mandate required the U.S. motor fuel supply to include 7.5 billion gallons of ethanol or other biofuels by 2012. This doubling of biofuels production over eight years was viewed as a sop to the powerful farm lobby by critics of the RFS who pointed out that it was a mere drop in the bucket compared to the 140 billion gallons of motor fuel consumed annually in the U.S. Many in the fuels industry were skeptical that agriculture could supply even this relatively small contribution to the U.S. energy supply. In fact, the RFS target was not only met but surpassed with ethanol production capacity reaching 13.6 billion gallons by the end of 2010.

More important than the RFS in stimulating the biofuels industry was Congress' refusal to provide liability protection to petroleum companies as they phased out MTBE manufacture. This put companies at risk of lawsuits if they continued to provide MTBE as an octane enhancer in gasoline. Unwilling to face the financial risk, petroleum companies very quickly shut-down MTBE production. The loss in volume and energy associated with MTBE increased the demand for higher grade blend stocks of gasoline, which could not be readily met with existing refining capacity in the United States. In anticipation of passage of the Energy Policy Act gasoline prices began to rise even while ethanol prices fell. By March of 2005 ethanol was cheaper per gallon than unleaded gasoline even before discounting for the fifty-one cent ethanol blender credit provided by Congress.

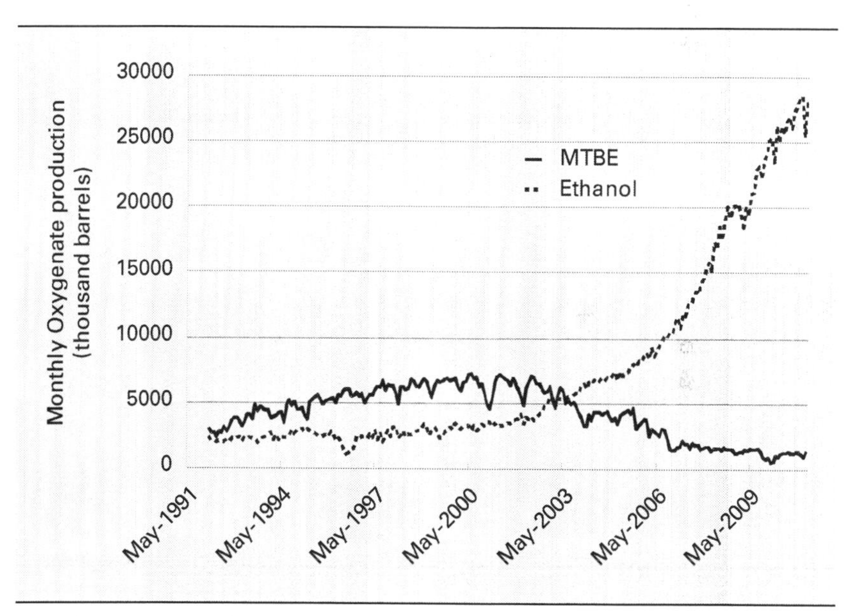

The rise and fall of MTBE. *Source: U.S. Energy Information Administration*

Adding to the attractiveness of ethanol were the profit margins for ethanol producers. A simple metric for grain ethanol profitability is the "corn crush spread," defined as the difference between the sales value of ethanol and the cost of corn grain, expressed as dollars per bushel. The crush spread does not account for operating costs or capital recovery costs for an ethanol plant, which varies widely in the industry, but it provides a useful proxy for the profitability of the ethanol industry. Corn crush spreads climbed from $1 per bushel in April 2005 to $5 per bushel in August 2005 and even briefly zoomed to $10 per bushel in 2006. For a time profits were as high as $1 per gallon of ethanol produced. These favorable economics launched a building boom that made the RFS and ethanol subsidy superfluous.

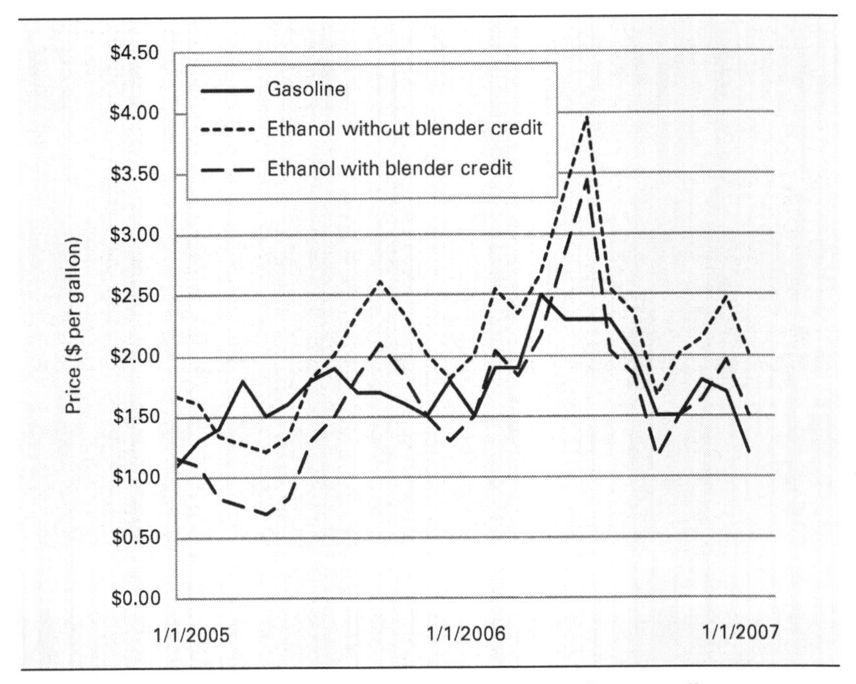

For a brief time in 2005 grain ethanol was cheaper than gasoline even without the blender credit. *Source: Energy Information Administration, Chicago Board of Trade, 2007*

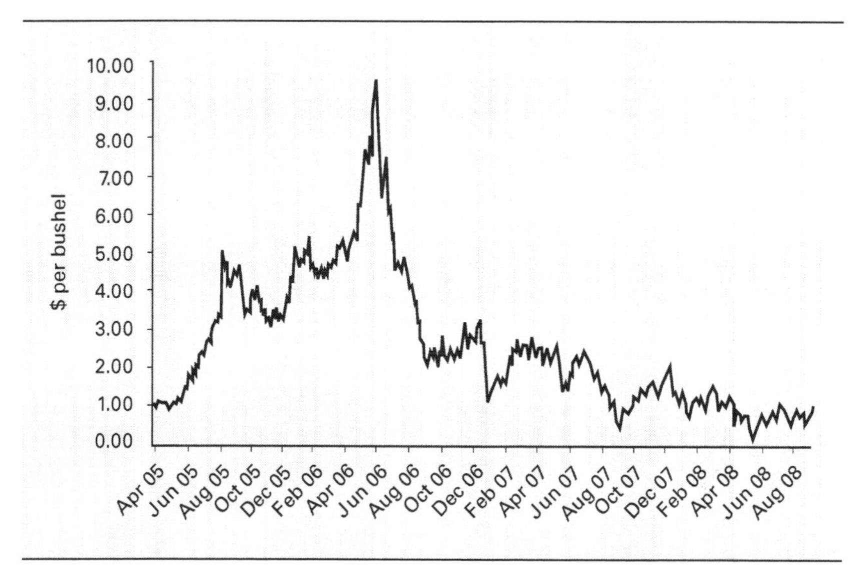

Attractive corn crush spreads in 2005 and 2006 contributed to the boom in biofuels. Crush spread is the estimated gross margin for ethanol production per bushel of corn, calculated according to the formula:
Crush Spread = 2.80 x Ethanol Price – Corn Price. *Source: Chicago Board of Trade, CBOT Ethanol Chartbook, September 2008*

The number of ethanol plants increased from a handful in 2004 to 173 plants by the end of 2008.

The future looked extremely bright for biofuels. New companies were incorporated to design, build, and operate plants to produce grain ethanol as well as biodiesel, an oxygenate from vegetable oils and waste animal fat that can substitute for diesel fuel in compression ignition (diesel) engines. Agricultural producers suddenly had a choice of markets for their crops. In the meantime corn prices, which had been hovering around $2 per bushel in early 2005, soared to $5.50 by the middle of 2008 and passed $6.00 in 2011. Farmers who provided both grain and capital to new plants

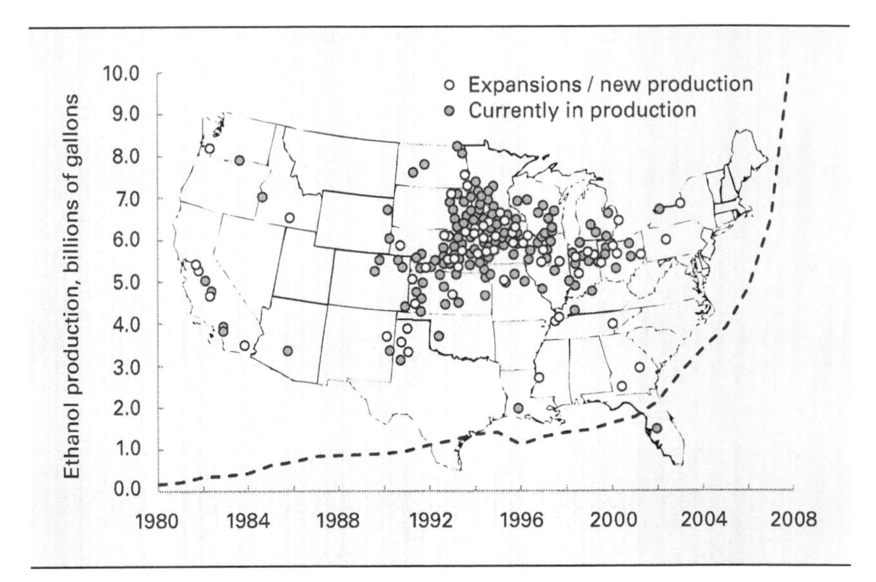

Renewable fuels in the United States expanded rapidly after the passage of the Energy Policy Act of 2005. *Source: Renewable Fuels Association (map based on mid-2008 statistics)*

found their financial prospects considerably improved from the 1980's when agriculture was so unprofitable that farm foreclosures swept across the land like a plague. Rural communities hoped that biofuels agriculture would rebuild their local economies, which had long been in decline.

Most importantly, the grain ethanol boom demonstrated that the United States could rapidly ramp up the biofuels industry at a time when concerns about national security and global climate change also dominated the national news. Domestic grain ethanol production is unlikely to ever provide the U.S. with more than about 15 billion gallons per year, which would displace around 10% of present or future demand for motor fuels. However, reaching half this production level was enough to alarm the Organization of

Petroleum Exporting Countries (OPEC). In June 2007 OPEC's secretary general Abdalla El-Badri told the Financial Times of London that moves to use biofuels would make his members consider cutting investment in new oil production.[3] Whereas just a few years earlier many observers had dismissed biofuels as a mere drop in the bucket in its ability to meet demand for motor fuels, OPEC's leadership was professing concerns about "security of demand" for petroleum as a result of the emerging bioeconomy.

Although grain ethanol alone cannot topple petroleum, it is the vanguard of dramatic changes coming to the energy industry in the form of advanced biofuels, a fact that even OPEC now recognizes. Others to recognize the magnitude of the coming changes are venture capitalists who have asserted that this so-called "clean tech" era will rival the information age in its impact on the way we do things. Venture capitalists invested $750

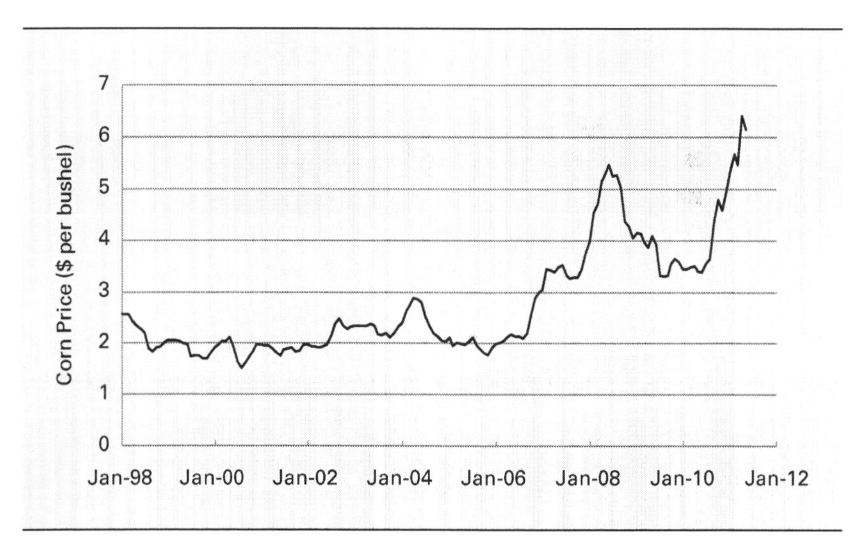

Expanding grain ethanol production helped drive up corn prices in 2007 and 2011. *Source: USDA National Agricultural Statistics Service.*

million in biofuels technology in 2007.[4] Advanced biofuels companies raised more than $450 million in IPOs during the first half of 2011 alone.[5] Prominent investors in biofuels include Vinod Khosla, Richard Branson, Bill Joy, and Bill Gates. Although some of these investments went toward grain ethanol plants, most were aimed at technologies that use cellulose or inexpensive sources of lipids in the production of biofuels ranging from ethanol to hydrocarbons.

The U.S. Congress acknowledged the limitations of grain ethanol and the importance of advanced biofuels in reducing demand for imported petroleum with the passage of the Energy Policy Act of 2005, which called for the blending of 7.5 billion gallons of renewable fuel with gasoline by 2012. This original Renewable Fuel Standard (RFS1) was updated and expanded by the Energy Independence and Security Act of 2007 (EISA).This legislation called for the annual production of 36 billion gallons of renewable fuel by 2022 as part of a new Renewable Fuel Standard (RFS2), which was implemented by the EPA in May 2009.[6] Of this total, only 15 billion could be corn (starch) based ethanol with the caveat that fuel from grain ethanol plants built after passage of EISA would have to demonstrate 20% reductions in life cycle greenhouse gas emissions compared to baseline petroleum-derived gasoline. The legislation also called for an additional 16 billion gallons of renewable fuels to be produced from cellulosic biomass such as wood or switchgrass with at least 60% lower life cycle greenhouse gas emissions than from gasoline. The remaining 5 billion gallons could come from other kinds of biomass including sugar cane and fats and oils as long as they achieved 50% reductions in life cycle greenhouse gas emissions.

Even before the RFS2 was implemented, criticism of biofuels intensified. For many years a few vocal critics had

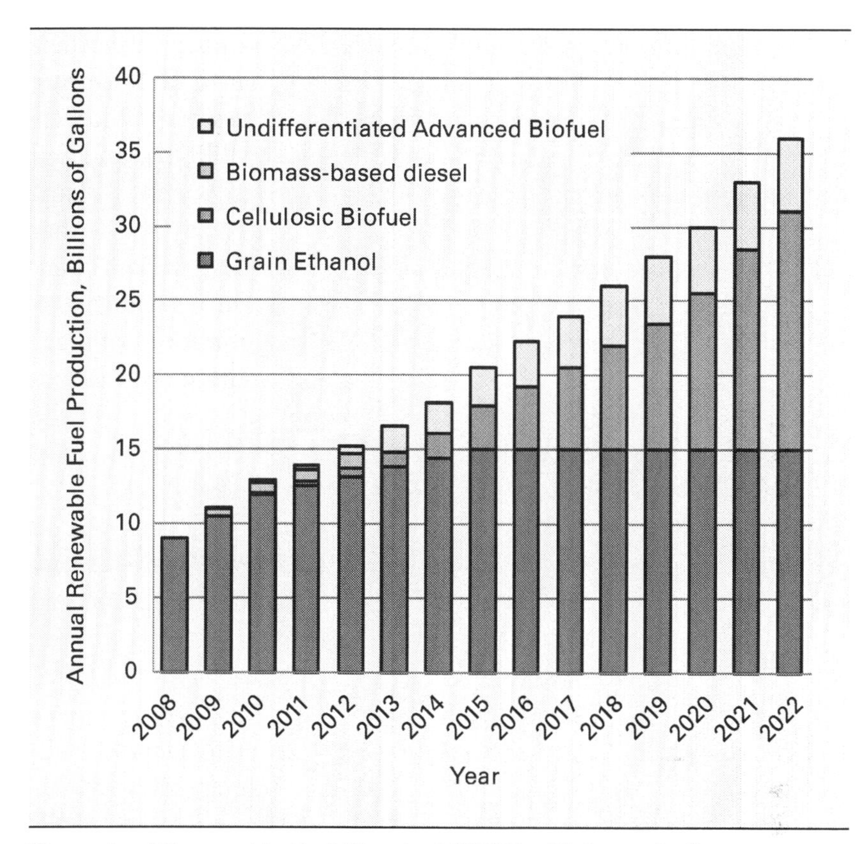

The revised Renewable Fuel Standard (RFS2) will dramatically ramp-up renewable fuels production in the United States. *Source: Renewable Fuels Association.*

persistantly claimed that ethanol consumed more energy in the form of fossil fuels during its production than was contained in the renewable fuel product. If true, then ethanol is unlikely to either save energy or reduce greenhouse gas emissions. Although this argument has several flaws (a point to be discussed in Chapter 7), it was widely circulated in the popular press, which raised doubts in many people's minds about the wisdom of ethanol as the center piece for the nation's energy future. These doubts were intensified by four developments that emerged in 2007 and 2008.

First, food prices began a run-up that corresponded with increasing prices for corn. Clearly, the demand for corn by the growing legions of ethanol plants was directly responsible for the increase in corn prices. In many people's minds the relationship between high food prices and high corn prices also seemed obvious although, as will be demonstrated in Chapter 7, it does not stand up to closer scrutiny. Some news reporters were quick to blame biofuels for everything ranging from the price of tortillas in Mexico to violent riots in southeast Asia over shortages of wheat and rice. Concerns that "food crops" were being diverted to fuels production at the expense of human nutrition escalated to the point that the United Nation's expert on food policy, Jean Ziegler, described biofuels as a "crime against humanity."[7] He called for a five-year moratorium on biofuel production to halt a "growing catastrophe for the poor." In April 2008, Scott Faber, the vice president of Federal Affairs at the Grocery Manufacturers Association, blamed the U.S. ethanol industry for rapidly rising food prices which he said were "devastating" for American families. Faber called on Congress to address this "unsustainable pattern" and "reexamine food-to-fuel mandates."[8]

Second, Timothy Searchinger, a visiting scholar at Princeton University and former attorney at the Environmental Defense Fund, and Joe Fargione, a researcher at the Nature Conservancy, published separate papers in Science on February 7, 2008 arguing that corn ethanol is responsible for land use changes in the developing world and hence, should be held responsible for the greenhouse gas emissions associated with these changes.[9,10] This so-called indirect land use change (ILUC) argument had been previously made by environmental groups, but these two papers were the first attempts to put numbers to the concept. Their publication received more than passing notice because they threatened the 36 billion gallon Renewable Fuel Standard passed by Congress and approved by the Bush Administration less than two months earlier. This legislation not only mandated an expanded renewable fuel standard but required new biofuels plants, whether processing starch, cellulose or other kinds of biomass, to reduce life cycle greenhouse gas emissions compared to fossil-based gasoline. The legislation included specific language defining "life cycle greenhouse gas emissions:"[11]

> The term "life cycle greenhouse gas emissions" means the aggregate quantity of greenhouse gas emissions (including direct emissions and indirect emissions such as significant emissions from land use changes), as determined by the Administrator, related to the full fuel life cycle, including all stages of fuel and feedstock production and distribution, from feed stock generation or extraction through the distribution and delivery and use of the finished fuel to the ultimate consumer, where the mass values for all greenhouse gases are adjusted to account for their relative global warming potential.

The California Air Resources Board (CARB) was already developing a "low carbon fuel standard" around the notion that being "renewable" was less important than being "low carbon." In principle, a renewable fuel does not emit net carbon dioxide because plants from which renewable fuels are produced consume as much carbon dioxide during their growth as is released from the production and utilization of the renewable fuel. In practice, fossil fuels are employed in growing, harvesting, and processing crops used for fuel. Furthermore, certain agriculture practices encourage the release of net carbon dioxide and other greenhouse gases from soils. These practices call into question the sustainability of renewable fuels from the standpoint of global climate change. Nevertheless, even accounting for greenhouse gas emissions "on the farm" and "in the manufacturing plant," renewable fuels are far superior to the fossil fuels they replace, which even biofuels critics concede. However, inclusion of so-called indirect land use change in the calculation of life cycle greenhouse gas emissions, as proposed by Searchinger and Fargione, leads to the paradoxical conclusion that fossil fuels are a better choice than renewable fuels with respect to efforts to slow global climate change.

The argument put forth is quite simple: if an acre of corn crop is diverted to biofuels production, then the resulting corn deficit in world markets will be filled by farmers converting rainforests and grasslands to agricultural lands. Depending upon the assumptions employed for this land conversion, the net carbon dioxide emissions potentially could overwhelm the emissions saved by using biofuels in place of gasoline. Both groups of researchers argued that this deficit, although not directly the result of biofuels agriculture, should be made the responsibility of ethanol producers. To many, this so-called indirect land use change argument seemed eminently reasonable in the face of a prospective greenhouse gas policy in the U.S. that only held certain sectors of the economy responsible for their carbon emissions.

The inclusion of indirect land use change in the definition of "lifecycle greenhouse gas emissions" of the Energy Independence and Security Act of 2007 was a coup for environmental groups opposed to ethanol who were otherwise unable to block the inclusion of the renewable fuels standard in the Energy bill. If the U.S. Environmental Protection Agency (the "Administrator" alluded to in the legislation) had calculated greenhouse gas emissions in the manner advocated by some environmental groups, then the era of biofuels, whether first-generation ethanol or advanced biofuels, would have been strangled shortly after its birth. As it was the final regulations ultimately adopted by the EPA found grain ethanol and advanced biofuels to have smaller carbon footprints than gasoline, provided certain energy-efficiency practices were adopted by producers.[12] CARB initially adopted regulations that attributed a higher carbon footprint to grain ethanol than gasoline but decided to re-evaluate its ILUC modeling after new studies found flaws with it.[13]

Of more immediate concern to the industry in 2008 was evidence that the grain ethanol boom had already expended itself, a victim of its overly-rapid expansion and the collapse of financial markets as the world entered a severe recession. Too many ethanol plants producing too much ethanol drove up the price of corn and other inputs while driving down the price of ethanol. By mid-2006 the ethanol crush spread had begun to drop from its peak of over $9 per bushel reaching less than $1 per bushel by late 2007. At one point in 2008 the crush spread dropped essentially to zero. The rising price of petroleum in 2008, reaching a record price of $147 per barrel in July, and expectations for a record corn harvest that fall gave hope to the biofuels industry for an improved outlook for ethanol. Instead, heavy spring rains and the floods that followed either reduced corn yields or completely wiped out corn crops in many parts of the Midwest. Corn futures continued to rocket upward and ethanol companies scrambled to secure contracts for corn supplies until corn prices peaked in July. At almost the same time that corn prices changed direction, markets for petroleum collapsed and gasoline fell from $4 per gallon to $1.50 per gallon in six months' time. Certainly the run-up in gasoline prices was well understood: increasing demand for energy, especially in China and India, was expected to put upward pressure on petroleum prices over the long term. The precipitous drop in gasoline prices was more difficult to understand but likely involved several factors including voluntary conservation efforts by consumers and the failure of OPEC to agree upon production limits; the rising value of the U.S. dollar, which decreased the domestic price of imported petroleum; the departure of speculators from petroleum trading, who had played a role in the ramp up of prices but were now fearful that the run-up was over; and certainly the collapse of financial markets in the fall of 2008, which tied up credit and reduced demand for products and services around the world.

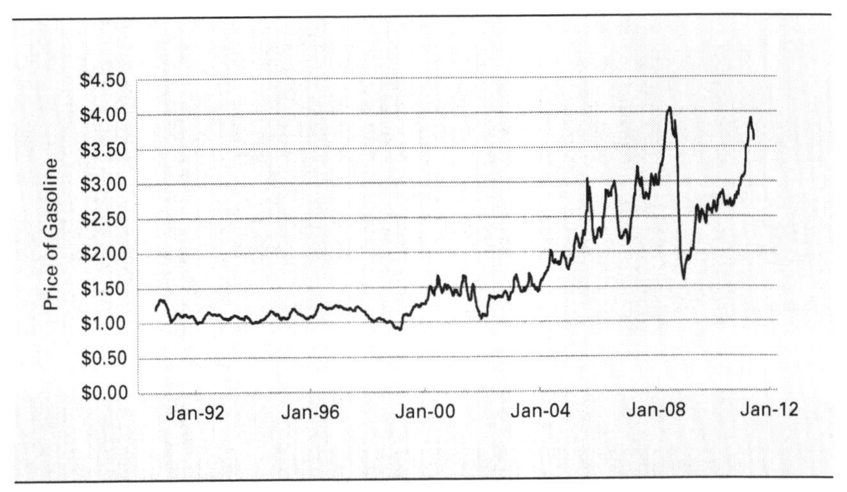

The rise and fall of the price of gasoline in the new millennium.
Source: U.S. DOE EIA

Finally, cellulosic ethanol, which is seen by policymakers as the successor to grain ethanol, has struggled to meet even a fraction of the cellulosic ethanol production targets set by the RFS2. Cellulosic ethanol has lower greenhouse gas emissions than grain ethanol and uses non-food feedstocks, allowing it to neatly sidestep two of the largest controversies surrounding grain ethanol. It can only serve as a successor to grain ethanol if produced in large enough quantities, however, and the magnitude of recent downward revisions to the RFS2 cellulosic ethanol mandate - 94% in 2010 to 6.5 MGY from 100 MGY, for example[14] - indicate that the industry is not yet prepared to replace grain ethanol as a major biofuel source. Multiple studies have found that the Department of Energy's (DOE) goal of producing cellulosic ethanol at a cost of $1.23/gal by 2012 (2010 dollars)[15] is still greatly out of reach with only one year remaining.[16, 17, 18] This raises serious concerns as to whether cellulosic ethanol will be capable of fulfilling its portion of the RFS2 in the foreseeable future.

Given the short duration of the grain ethanol boom, it is fair to ask whether the biofuels industry has run its course and whether it is time to turn to more promising sources of energy. But it should be remembered that the end of a boom is not necessarily the end of an era. A boom frequently presages a period of more sustainable economic growth. America's dotcom boom of the 1990's crashed spectacularly but spawned the era of the knowledge worker; the Florida land boom of the 1920's went bust but set the state on a course of long-term economic and population growth; and no one disputes that from the California gold rush emerged a state with an economy that rivals the gross domestic products of most nations of the world. The question of whether biofuels deserves a similar legacy is explored in subsequent chapters.

3

Why do we need to end our addiction to oil?

We live in a petroleum economy. Petroleum not only provides 35% of primary energy production in the world,[1] it is the basis of modern society, providing fuel for most of our transportation infrastructure as well as the chemical building blocks for many industrial applications and consumer products. Petroleum has made possible the automobiles, aircraft, synthetic fibers, personal electronics, inexpensive consumer goods, health care delivery, and convenience foods that characterize modern society. We would be lost without petroleum. Yet at the same time we need to end our "addiction to oil." The petroleum economy, despite the many benefits it has brought to society, is not sustainable. In the short term, U.S. dependence upon it imperils future economic growth. In the longer term, world supplies of petroleum are not sufficient to support growing world demand. But even before we "run out of oil" there is concern that the wholesale transfer of carbon from subterranean petroleum deposits to the atmosphere when it is burned as transportation fuels will disturb the Earth's climate in ways that might negate the benefits of the petroleum economy. This chapter examines the reasons for ending our addiction to oil.

U.S. production cannot meet U.S. demand

In 1956 a geologist for Shell Oil named M. King Hubbert predicted that U.S. oil production would peak in the early 1970s.[2] His prediction, now known as "Hubbert's Peak," was widely derided until 1970 when U.S. oil production began falling. With the brief exceptions of the twin oil shocks of the following decade, it has continued to do so ever since.

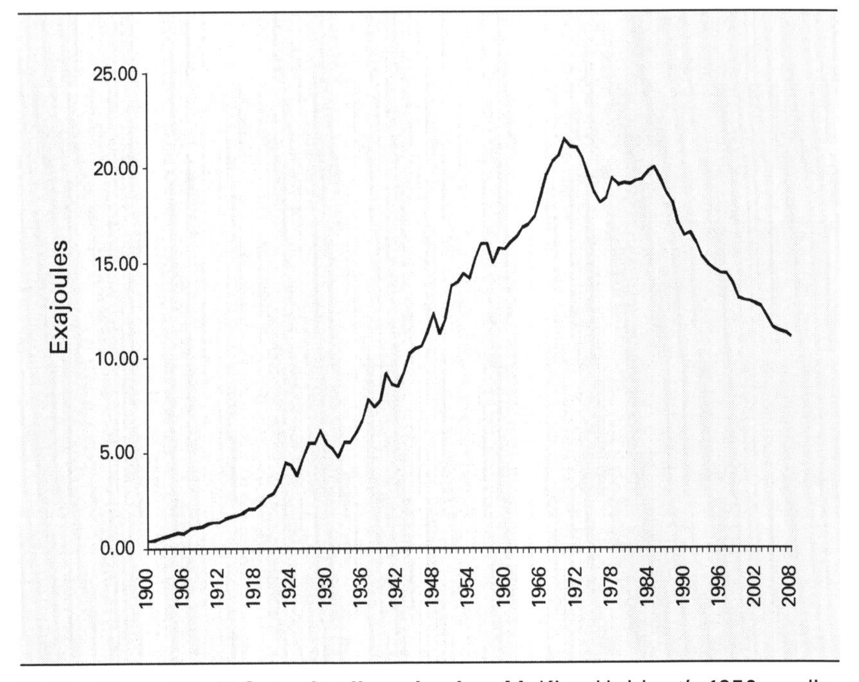

Hubbert's peak in U.S. crude oil production. M. King Hubbert's 1956 prediction that U.S. oil production would peak in the early 1970s is now a fact. *Source: EIA, BP.*

This steady drop in U.S. production has been matched in rapidity by the rise of U.S. oil consumption. U.S. production was equal to 67% of U.S. oil consumption in 1965, 43% in 1990, and a mere 25% by 2008.[3] The deficit has been made up by imports from other countries. Imports increased rapidly after peak oil was reached in 1970, declining after the Oil Crisis of 1979 in response to strict conservation efforts, but again increasing as the crisis eased in the mid-1980s. By 1994 our reliance on petroleum imports exceeded our domestic production of petroleum. The threat of overreliance on imported petroleum to energy security is clear from the historical record. U.S. oil imports took noticeable hits during World War II, the 1973 OPEC Embargo, the 1979 Oil Crisis, and the 1st Gulf War. Should a similar drop in imports occur now that the bulk of U.S. oil consumption is derived from foreign sources, the impact on the U.S. economy would be serious.

The United States' North American Free Trade Agreement (NAFTA) partners, Mexico and Canada, are its two largest oil suppliers, with the U.S. importing 23% of its supply from just these two countries.[4] Nearly half of the United States' remaining oil imports come from countries that are either engaged in violent civil strife or have adopted hostile stances towards the U.S. These countries, in order of the amount of petroleum imported to the U.S, are Nigeria, Saudi Arabia, Venezuela, Angola, Iraq, and Columbia. Finally, 53% of American oil imports come from member nations of the Organization of Petroleum Exporting Countries (OPEC), a cartel dedicated to keeping the price of crude oil artificially high for its economic gain by carefully controlling output. OPEC has demonstrated its willingness and ability to use production controls for political purposes, as well. The 1973 OPEC Embargo was launched for the sole purpose of forcing a change in U.S. foreign policy in the Mideast, which was perceived to favor Israel over its Arab neighbors.

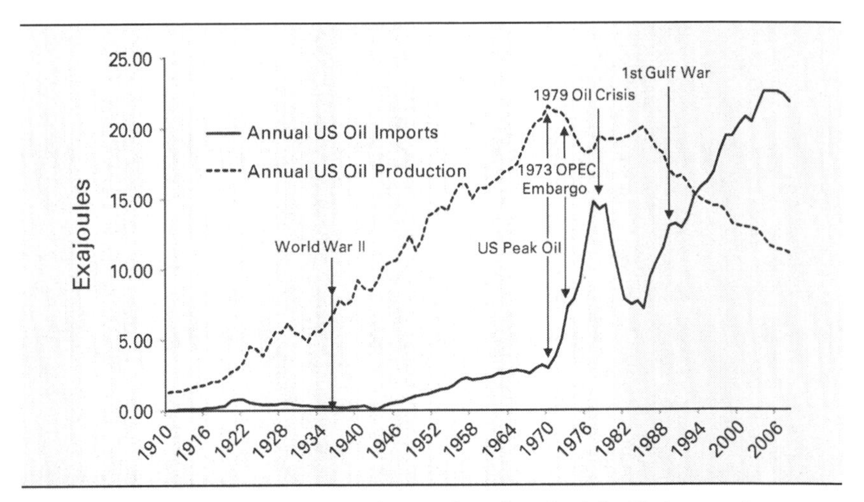

Petroleum imports drop during times of national crisis. Today we import more petroleum than we produce domestically, which will exacerbate the problem of energy supply during the next national crisis (petroleum production and imports expressed as exajoules of energy). *Source: EIA*

World supply will be hard pressed to meet U.S. demand

Hubbert's method of analyzing historical data on oil discovery and production to predict future supply, first applied to the United States petroleum resource, has also been able to predict peak oil for other countries. In 2004 the U.S. government reported that most oil-producing countries outside OPEC and the former Soviet republics have indeed experienced peak oil.[5] In 2005 the Worldwatch Institute found that oil production had begun declining in 33 of the 48 largest oil-producing countries.[6] These successful applications of Hubbert's method suggest its use as a tool in predicting the time when the world will "run out of oil."

But predicting peak oil for world petroleum production is not as simple as predicting peaks for individual countries for

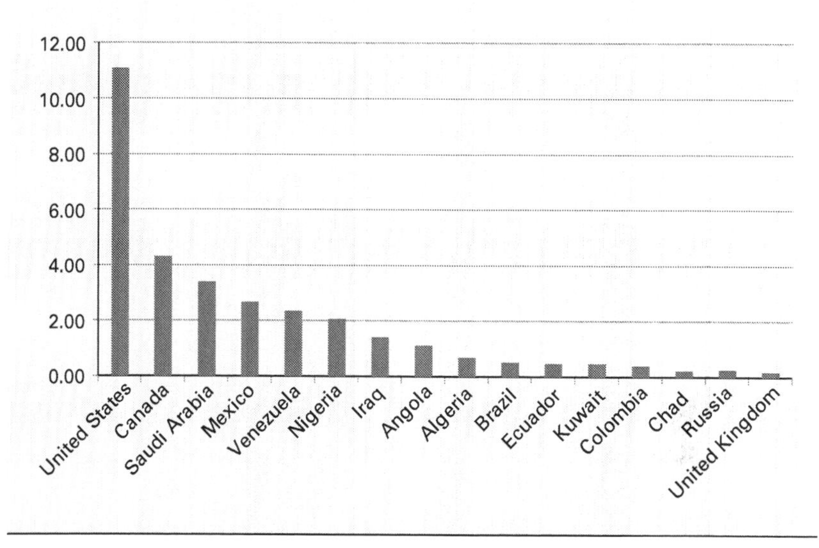

The U.S. gets its oil from diverse and not always reliable trading partners (imported petroleum expressed as exajoules of energy). *Source: EIA*

a number of reasons. Calculations for individual countries were based in part on oil reserve figures that were relatively well known. Accurate world reserve figures are harder to come by. In the 1980s OPEC member nations revised their oil reserve figures substantially upwards, resulting in numbers that geologist Kenneth Deffeyes has described as "an overly optimistic view of future oil production."[7] Less generously, energy expert Vaclav Smil characterized these self-serving projections as a "prime example of politically generated reserves."[8] OPEC members have a substantial financial interest in convincing their customers that the world oil supply is in little danger of being exhausted, thus discouraging attempts to develop alternative fuels. It is unlikely to be a coincidence that the President and CEO of Saudi Aramco, which alone is responsible for nearly 10% of the entire world's oil production, stated in 2008 that only 6-8% of the world's total oil reserves have been extracted thus far.[9] Saudi Aramco's upward revision greatly expanded the world's declared reserves and put in doubt earlier attempts to predict world peak oil.

Finally, estimating peak oil requires accurate information on world-wide petroleum consumption. While it seems certain that petroleum consumption will grow rapidly as the populations and economies of developing countries like India and China expand, their actual consumption rates will depend upon a number of uncertain factors. Will increasing education and public policy reduce the rate of population growth? Will the rising middle class in these countries have the same expectations for high protein diets and personal transportation as in the West? Will China and India join international climate change efforts? Given that forecasters are faced with questionable reserves estimates, changing technology, and widely disparate population and economic growth projections upon which world peak oil prediction must rely, their job relies as much on guess work as it does on rigorous analysis.

Hubbert himself predicted that world peak oil would occur in 2003 or 2004,[7] a projection that was clearly premature. There has been no shortage of other attempts to predict world peak oil, several of them falling within the first decade of the twenty-first century while others extend several decades hence.[10]

This uncertainty should not be viewed as justification for complacency about America's energy security. At current rates of petroleum extraction it is evident that the conventional reserves of Mexico and Canada will be exhausted within 11 years and 26 years, respectively, and nearly the entire world's proved conventional reserves will run dry

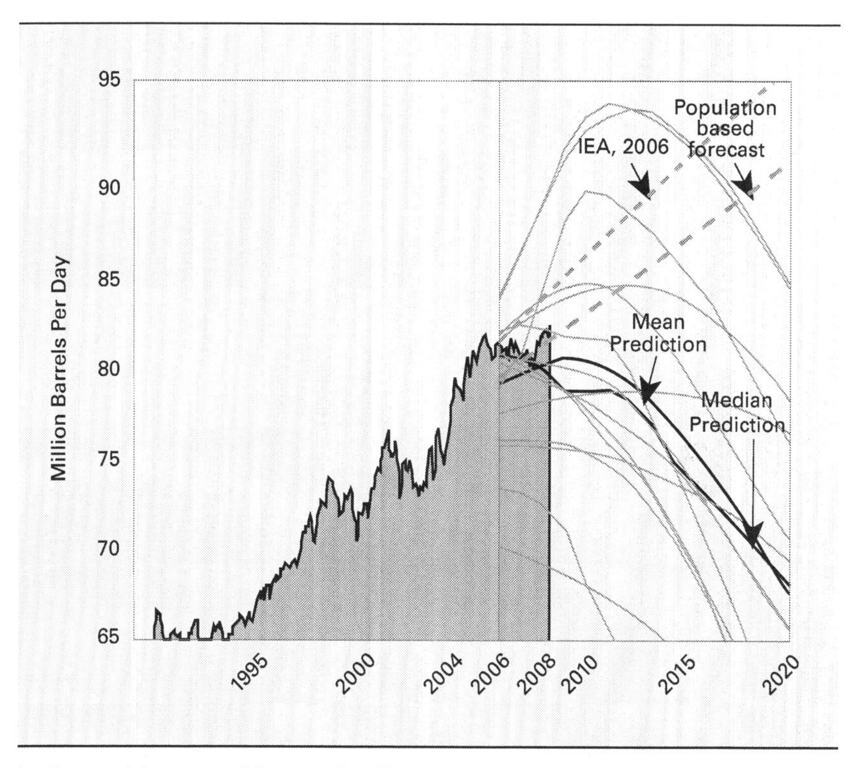

Is the world approaching peak oil? It depends upon who you ask. *Source: Redrawn from Reference 10.*

within 46 years (these figures do not include probable reserves and unconventional sources, which are among the alternative energy resources described in Chapter 4).[11] In 2010 the International Energy Agency recognized the existence of global peak oil for the first time by calculating that conventional petroleum production peaked in 2006.[12] While this doesn't take into account unconventional petroleum reserves, these have cost and environmental issues of their own that are discussed in Chapter 4. Of course, petroleum reserves will never be completely exhausted because prices of the increasingly scarce resource will become so high as to encourage markets to switch to alternative energy resources.

While reserves predictions tend to be conservative,[13] these remain sobering statistics. Even Iran, which currently sits on 11.2% of the world's oil reserves, giving it control over the world's second largest oil reserves, has ambitions for domestic biofuels production using excess crops.[14] In the meantime, the world finds itself increasingly reliant upon OPEC member nations which have a vested interest in manipulating oil prices and the means to do so. Furthermore, of the three countries with the world's largest oil reserves, one is governed by religious extremists (Saudi Arabia), one has a lengthy history of open hostility to the U.S. and is currently under a trade embargo (Iran), and the third has had difficulty maintaining steady petroleum production because of years of violent sectarian strife (Iraq).

Petroleum is a major source of greenhouse gas emissions

That atmospheric carbon dioxide plays a crucial role in determining climate and maintaining life on Earth is well established. It is also clear that burning petroleum-derived fuels and other fossil fuels contributes to the burden of carbon dioxide in the atmosphere, which has increased from 315 ppm to 383 ppm over the last fifty years, an increase of 22%.[16] More difficult to ascertain is whether the average global temperature is increasing as a result of human activity, which computer models of global climate suggest should be the case. Even more difficult to answer is whether continued reliance on fossil fuels could so alter the climate as to make the Earth a less hospitable place for its inhabitants.

While the impact of greenhouse gas (GHG) emissions on

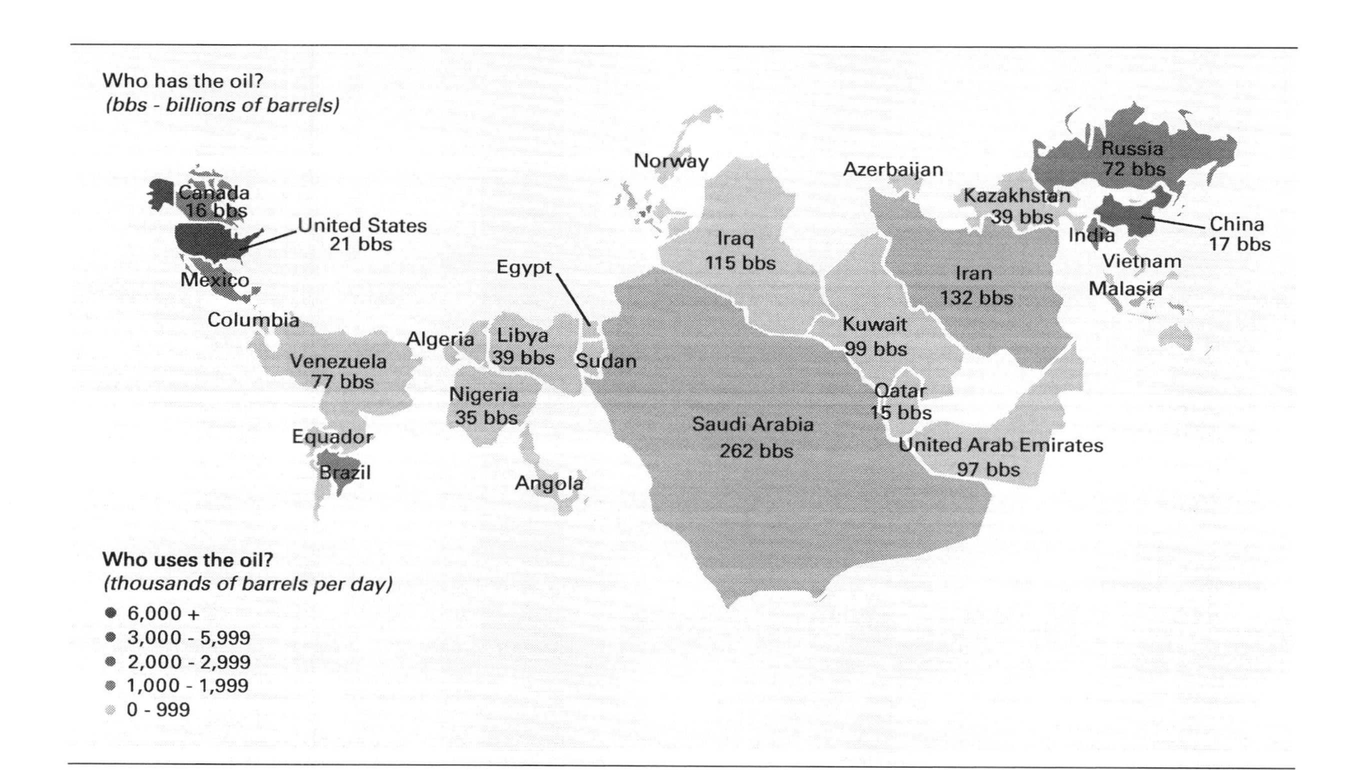

Who has the oil? Global map scaled according to the size of petroleum resources around the world. *Source: Redrawn from the BP Statistical Review Year-End 2004, which is based on the CIA's World Factbook.*[15]

global warming was an oft-contested point throughout the opening decade of the twenty-first century, a growing consensus within the scientific community holds that anthropogenic emissions of CO_2, CH_4, ozone (O_3) , N_2O, and chlorofluorocarbons (CFC) are resulting in the rapid warming of our planet.[17] A study published in the Proceedings of the National Academy of Sciences in early 2009 added urgency to the situation by predicting global warming will last for the next thousand years even if GHG emissions are reduced immediately.[18] With the carbon emissions of developing countries expected to increase drastically in the near future as the internal combustion engine becomes more prevalent in India and China, any attempt to decrease global emissions significantly will require the support of all nations if it is to succeed.

A range of maximum acceptable CO_2 concentrations has been suggested. Climatologist Wallace Broecker has proposed 560 ppm.[19] The Intergovernmental Panel on Climate Change recommends no more than 500 ppm.[20] NASA has reported that even 450 ppm may be unacceptable.[21]

It is estimated that the transportation sector (which uses petroleum-based fuels for all but a fraction of its energy needs) is the single largest source of carbon dioxide emissions and is responsible for nearly one-third of the world's emissions of this greenhouse gas.[22] Clearly, reducing atmospheric concentrations of carbon dioxide will require major reductions in the world-wide use of petroleum.

We no longer can afford to live in a petroleum economy

The 2008 financial crisis and subsequent global recession revealed an additional danger in relying on foreign sources of petroleum. Several notable economists have pointed to

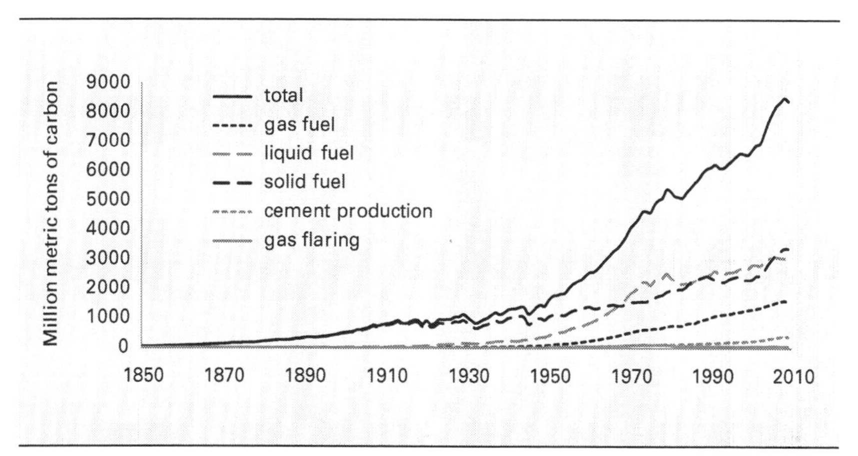

Petroleum is the leading contributor to fossil fuel emissions of carbon.
Source: Reference 22.

the huge U.S. trade deficit as one of the underlying causes of the crisis.[23, 24, 25] As petroleum prices reached historical highs in 2006-2008, petroleum-exporting countries were unable to spend all of their petroleum revenues and began searching for safe investments. They had to look no further than the U.S. government, historically one of the most reliable debtors in the world, which was in need of lenders to finance its burgeoning debt. To quote The Economist, "the money flooding in from willing foreign savers had bid up government bond prices, lowering interest rates and lifting house prices."[26] The housing boom in turn spurred the development and proliferation of complex and opaque financial instruments based on commercial and residential real estate. The violent end to the housing boom caused a corresponding collapse in the value of these instruments and the liquidity of the numerous major financial institutions that were overexposed to them, launching an era of high unemployment and slow economic growth that many countries, including the U.S., have yet to recover from. This is not to say that America's thirst for petroleum was entirely responsible for the economic collapse; China, also a petroleum importer, also invested heavily in U.S. debt prior to the financial crisis. A cautious response by the world financial industry – as opposed to its headlong plunge into the "easy credit" markets – would have helped blunt the impact of the housing bubble's implosion. This doesn't change the inescapable fact that America's pain at the pump in 2006 and 2007 greatly contributed to the surge of pink slips and shrunken retirement accounts in 2008 and 2009.

Many economists have also argued that the oil-fueled U.S. trade deficit also presents a barrier to full economic recovery in the U.S.[27, 28] Policymakers have also placed an

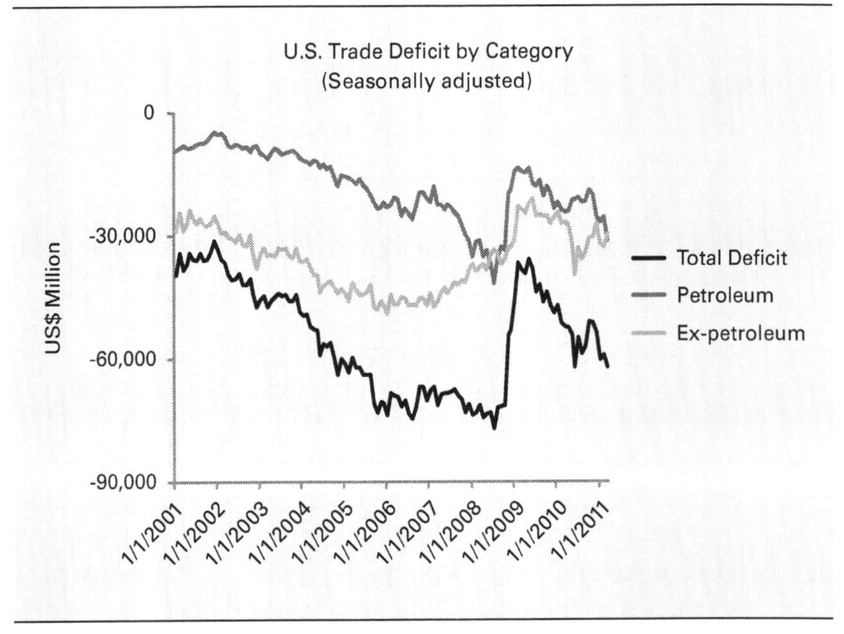

The U.S. trade deficit by category since 2001. *Source: U.S. Census Bureau.*

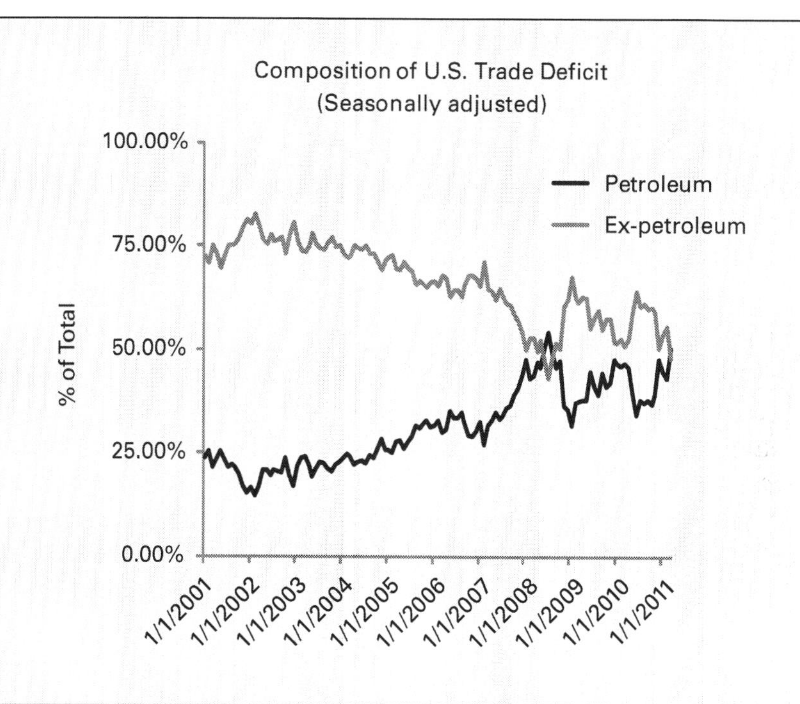

Composition of the U.S. trade deficit since 2001. *Source: U.S. Census Bureau.*

emphasis on a global rebalancing, as evidenced by President Obama's twin declared goals of achieving more favorable trade relations with China (i.e., increasing the price of Chinese goods imported to the U.S.) and a doubling of U.S. exports by 2015.[29, 30] A more detailed examination of the U.S. trade deficit suggests that such a rebalancing can instead be achieved by reducing U.S. oil consumption. For only the second time this century, the majority of the deficit is the result of petroleum imports.

This illustrates the importance of America's petroleum dependency to its trade deficit and the difficulty of a rebalancing as long as it continues. Indeed, the prospect of a painfully slow U.S. economic recovery has concentrated the minds of policymakers (and their constituents) to focus on the global rebalancing necessary for continued economic growth. This new focus on the economy does not have to necessarily come at the expense of environmental concerns, provided both can be addressed by the same remedy.

How about just reducing energy use?

Rather than find alternatives to petroleum to fuel our economy, some would argue that we need merely reduce our energy consumption through energy conservation and energy efficiency measures. Energy conservation is measures that reduce end-use demand for energy by reducing the service demanded. Examples are turning down the thermostat in homes and offices during the winter and up in the summer; turning off electric lights in unoccupied rooms; and carpooling so that fewer vehicles are on the road. Energy conservation has a connotation of giving up some level of comfort or convenience to which we have grown accustomed. Energy efficiency is measures that provide the same level of services but with less energy consumed. Examples are high efficiency furnaces; fluorescent lighting; and hybrid-electric vehicles. Energy efficiency would seem to involve less sacrifice and more reliance on technology to reduce energy use although at a higher cost than the common sense approaches of energy conservation. Sometimes it is difficult to distinguish between these two approaches to saving energy: does driving a smaller automobile that gets higher fuel economy at the expense of a less spacious interior represent an energy conservation measure or an improved energy efficiency measure?

Clearly, if we were to double the average fuel economy of the U.S. automobile fleet, we could halve the amount of gasoline consumed in the U.S., potentially reducing the need to import petroleum by 2.5 billion barrels per year. But how would this scenario really play out? Such a dramatic reduction in demand for transportation fuels would have the unintended consequence of reducing the price of gasoline. Although this would be welcomed by consumers, they would respond by driving more and buying larger vehicles, which would reduce the actual fuel savings expected from doubling fuel economy of automobiles. Even if the price of gasoline were to remain constant, the annual fuel costs of consumers would halve, encouraging them to make choices that would reduce the theoretical savings. This is what Peter Huber and Mark Mills[31] refer to as the "efficiency paradox:" individual consumers and companies alike will sacrifice energy efficiency for other kinds of efficiency, such as saving time or money, if energy costs remain low enough.

In fact, Huber and Mills argue that the economy cannot grow without increasing energy consumption – energy conservation and even energy efficiency are counterproductive to expanding economic opportunities. The history of the U.S. economy does not support this part of their thesis. Since the U.S. government began collecting energy statistics in 1949, the U.S. economy, as measured by the gross domestic product (GDP) in real dollars (that is, adjusted for inflation), has been expanding without significant pause. We might expect a similar trend for U.S. energy consumption. Up to 1974 there is a correlation between U.S. energy consumption and U.S. GDP but then over a period of ten years, energy consumption stagnated and even dropped. In 1983, energy consumption was no greater than it had been in 1973, but the economy had grown by 25% in real dollars. The origin of this reversal, of course, was the oil embargo of 1973 by the Organization of Petroleum Exporting Countries (OPEC). Energy conservation and energy efficiency measures put in place in response to this disruption in petroleum supplies were effective in reducing energy demand in the U.S. while allowing the economy to grow. Although energy consumption began to grow again after 1983, the economy had fundamentally changed—energy intensity was dropping. Energy intensity is the amount of energy consumed, measured in British thermal units (Btu), per real dollar of gross domestic product. Whereas the energy intensity of the U.S. economy averaged 18 Btu per dollar between 1949 and 1974, it dropped to 11 Btu per dollar between 1983 and 2007, decreasing virtually every year in that period.

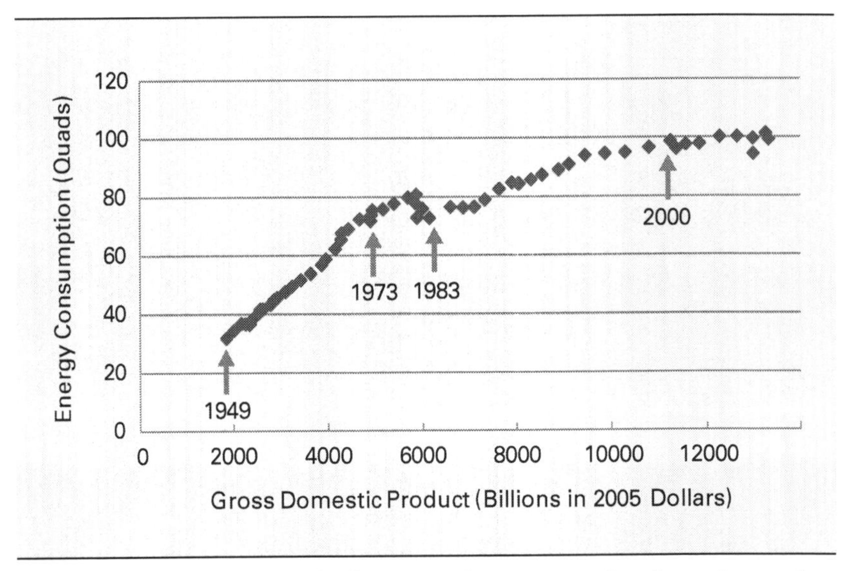

The amount of energy required to grow the economy has been decreasing since the oil shock of 1973. *Source: Energy Information Administration.*

There are several reasons why the economy had grown more efficient in its use of energy. In addition to long-term implementation of energy conservation and energy efficiency measures, energy intensive industries like steel production moved overseas, replaced with less energy intensive service industries. Thus, economic growth may correlate with increasing energy consumption, but the amount of energy required per unit of growth can be dramatically decreased through energy conservation and energy efficiency. Indeed, as we move away from an era of inexpensive, easily extracted fossil fuels it will be important to select energy systems that yield the most "bang for the buck" in terms of the miles driven or kilowatt-hours of electric power obtained from alternative energy sources.

4

What are our alternatives to imported petroleum?

In discussing the alternatives to imported petroleum, we must distinguish between primary energy sources and energy carriers. Primary energy sources are forms of energy as they are found in the natural environment. Rarely are primary energy sources in forms that can be conveniently utilized. Petroleum, for example, is a primary energy source incorporated into geological strata as a viscous mixture of liquid hydrocarbons. Even sweet, light crude oil is not ready to put into an automobile without chemical upgrading. In contrast, energy carriers are convenient and "clean-burning" forms of energy that are readily transported, stored, and utilized in distributed applications. Nature rarely provides us with energy in a form that meets our modern expectations for energy carriers and we resort to physically or chemically transforming primary energy sources into suitable energy carriers. Gasoline, for example, is an energy carrier manufactured from petroleum.

Early technologies often attempted to utilize primary energy sources as energy carriers. For example, wood and coal were successfully used as locomotive fuel in the early years of the railroad era. As we shall see, this almost always gave way to some kind of chemical upgrading or energy conversion to yield an energy form more useful than the energy form found in nature. Sometimes the atoms or molecules

We must choose among both primary energy sources and energy carriers to devise alternatives to petroleum-based transportation fuels.

Primary energy sources	Energy carriers
Coal	Hydrocarbons
Petroleum	Hydrogen
Natural Gas	Methane
Tar sands	Ammonia
Oil shales	Dimethyl ether
Methane hydrates	Ethanol
Nuclear	Methanol
Wind	Butanol
Hydropower and wave energy	Fischer-Tropsch Liquids
Geothermal	Furans
Tidal energy	Esters
Solar electric	Terpenes
Photosynthesis (biomass)	Electricity

that constitute a primary energy source are transferred to the energy carrier. Certainly this is the case when petroleum is converted into gasoline. On the other hand, when coal is used to produce hydrogen fuel, none of the carbon in the primary energy source ends up in the energy carrier. Oxidizing the carbon in the coal to carbon dioxide releases the energy needed to decompose water into molecular hydrogen. If the carbon dioxide can be kept out of the atmosphere by sequestering it into deep ocean waters or geological formations, then the energy in the coal has been recovered as a carbon-free hydrogen fuel.

Finding an alternative primary energy source to replace petroleum does not necessarily mean adopting new kinds of energy carriers. In principle, gasoline that is essentially identical to the mixture of hydrocarbons currently burned in automobiles could be manufactured from coal or biomass. Whether we stick to gasoline or switch to hydrogen, for example, depends upon a wide variety of factors including the nature of the alternative primary energy source used to produce the energy carrier, technical limitations, environmental considerations, political constraints, and economic drivers. Thus, a familiarity with the various alternative energy sources under consideration is an important part of the discussion of alternatives to imported petroleum.

Alternatives to imported petroleum as primary energy sources include other fossil fuels, nuclear power, and renewable energy. The fossil fuel options include domestically produced petroleum, natural gas, coal, and unconventional fossil fuels, which include tar sands, oil shale, and methane hydrates. Nuclear includes both fission reactors and fusion reactors, although the latter is unlikely to play a role until well after the middle of the twenty-first century.[1] Renewable fuels include wind power, water-in-motion (hydroelectric, tides, and waves), geothermal, solar electric, and photosynthesis (producing biomass both natural and synthetic).

Domestically produced petroleum

As the previous chapter made clear, the United States is no longer a heavyweight when it comes to petroleum reserves. Nevertheless, we have reserves of 21 billion barrels of petroleum that could be developed relatively quickly compared to other primary energy sources.[2] The difficulty is that over 50% of these reserves are in Alaska, California, and off-shore locations where there is considerable political resistance to developing them either because they are near population centers or their exploitation might affect environmentally sensitive areas. The massive oil spill that occurred in the Gulf of Mexico in 2010 following the destruction of the *Deepwater Horizon* oil rig and subsequent moratorium on offshore drilling in the Gulf highlighted the difficulties encountered by efforts to increase domestic petroleum production. Thus, there has been considerable enthusiasm over the recent announcement that the estimated reserves for the Bakken Formation in North Dakota and Montana has increased from 150 million barrels to as much as 4.1 billion barrels, which would make it the largest U.S. oil reserve outside of Texas.[3] Since much of the Bakken Formation occurs in relatively remote ranch land, there is little resistance to expanded oil drilling in the region. Off-setting the enthusiasm for this new, large reserve is the sobering

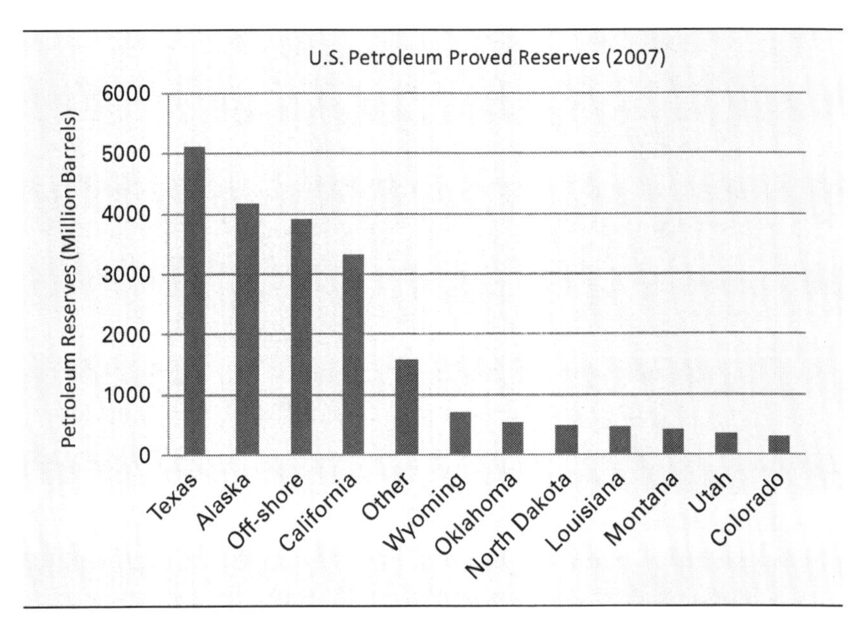

Much of the U.S. petroleum reserves exist in parts of the country where there is considerable political opposition to developing natural resources.
Source: U.S. DOE EIA

recognition that it represents at most a six month supply of gasoline at current rates of consumption in the United States. We need to look beyond domestically produced petroleum for future energy supplies.

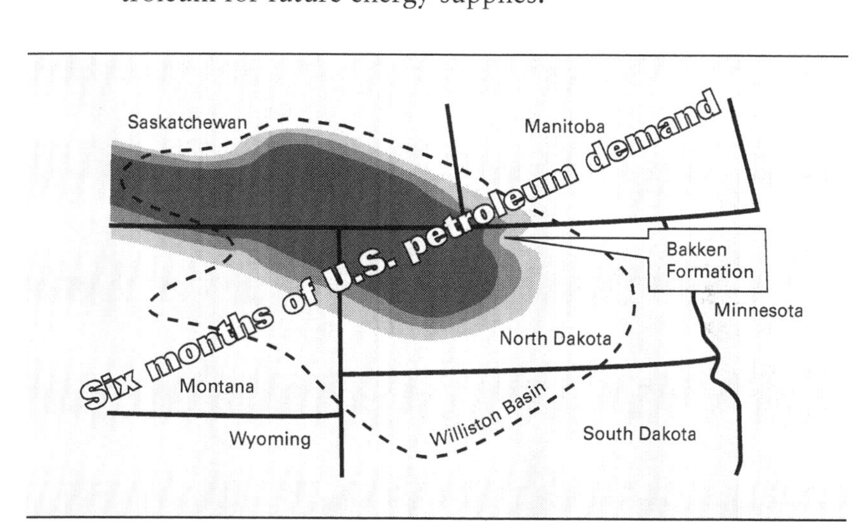

Drop in the bucket. The Bakken Formation although potentially the second largest petroleum reserve in the United States, only represents six months supply of gasoline for the U.S. *Source: Reference 3.*

Natural gas

Natural gas is a gaseous form of fossil energy, consisting mostly of methane, found in several kinds of geological deposits.[4] The "conventional sources" of natural gas are associated with oil fields or in so-called "non-associated natural gas fields." These were among the first to be developed either because they were a by-product of petroleum extraction or they were otherwise easy to exploit. Unconventional natural gas includes gas associated with tight sand formations, coal beds, and shale deposits. These are usually more difficult to develop than conventional sources but are often very plentiful and widely distributed. Whatever the source, natural gas as pumped out of the ground is not suitable for direct use because of the presence of non-combustible gases, water and large molecular weight hydrocarbons that can condense in pipelines and contaminants like sulfur that become air pollutants when the gas is burned. After removal of these undesirable compounds, the refined product is marketed as natural gas although for all intents and purposes it is methane (typically greater than 95% by volume). Refined natural gas is a convenient, clean-burning fuel that is readily transported by pipeline. Because it contains less carbon than other fossil fuels, substitution of natural gas for petroleum and coal could also help reduce the rate of greenhouse gas emissions although it certainly is not greenhouse gas neutral. Although methane is a more potent greenhouse gas than carbon dioxide, studies have indicated that fugitive emissions from increased natural gas would be a relatively small contributor to global climate change. Refined natural gas can serve as an energy carrier. The financier T. Boone Pickens believes natural gas is too valuable as transportation fuel to burn it in power plants. He advocates the use of wind energy to generate electric power, freeing up the use of compressed natural gas as transportation fuel.[5]

Thus, the question is how much natural gas is available to the United States for the production of transportation fuels? Proved reserves within the United States, including off-shore, are 237 trillion cubic feet of natural gas. We are currently consuming natural gas at the rate of 23 trillion cubic feet per year. If we were to also use this to replace our current demand for gasoline, our proved reserves of natural gas would be gone in less than six years. Even T. Boone Pickens plan of using wind power to save natural gas for transportation fuel would only extend this supply by one year since only 28% of natural gas is currently used for electric power generation.

But limiting the supply to proved reserves is too conservative an estimate. Proved reserves as defined by the U.S. DOE Energy Information Administration are those reserves for which there is a reasonable certainty that the natural gas recovered will meet or exceed the estimate. In fact, there are also "unproved reserves" of natural gas that are difficult to estimate accurately without further exploration and development of the resources. Many of these unproved reserves fall into the category of unconventional natural gas. A recent study suggests that unproved reserves are ten times larger than the proved reserves.[6] If correct, the total natural gas

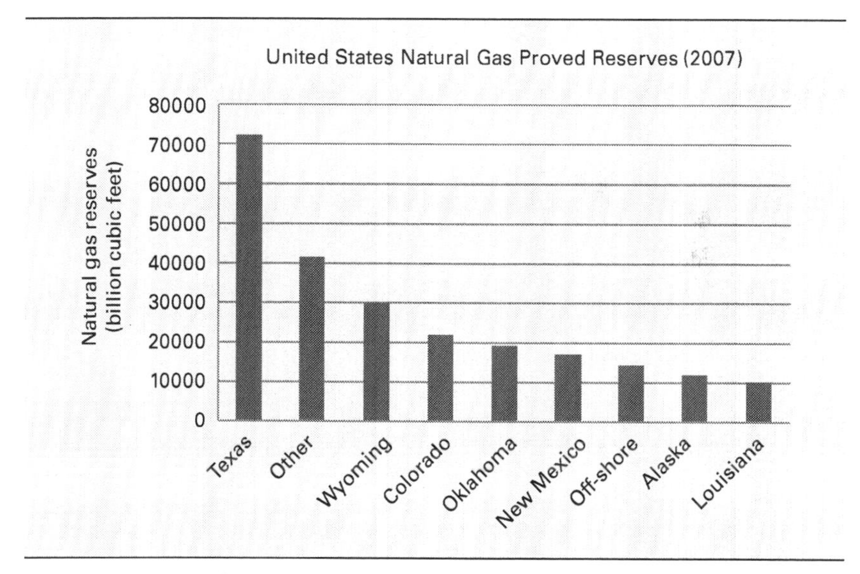

Proved reserves of natural gas in the United States could supply current rates of natural gas and gasoline consumption for less than six years.
Source: U.S. DOE EIA.

supply in the U.S. could be as large as 2,250 trillion cubic feet. This could supply our current natural gas and gasoline demand for 54 years. These unproved natural gas reserves are more difficult to recover. Much of the unproved reserves are trapped in dense deposits of shale, which require advanced recovery techniques. In a process known as "fracking," water and chemicals are injected into shale deposits to fracture the rock, which releases the trapped methane. Its rapid deployment in recent years has raised concerns about ground water contamination.[7]

Another concern is the historical volatility of natural gas prices. The principal drivers behind this volatility are supply and demand fundamentals, which include the weather, storage activities, and the perception of market conditions.[8]

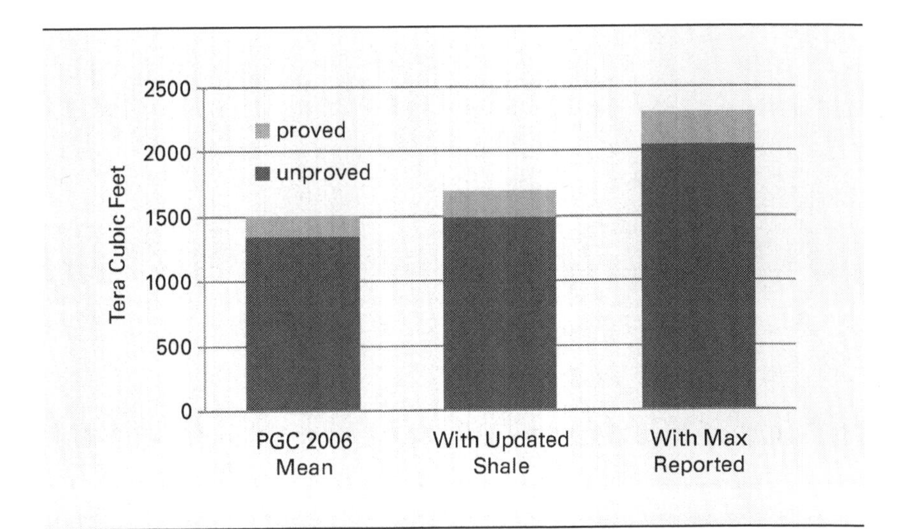

Unproved sources of natural gas in the United States dwarf proved reserves. It might supply traditional natural gas markets and serve as a substitute for gasoline for up to 54 years at current rates of demand. *Source: Reference 6.*

Among the strongest short-term drivers on the demand side is simply the forecast of seasonal temperatures. Nearly 40% of natural gas consumption comes from residential and commercial customers, much of it for heating of homes and businesses.[9] As a result, natural gas has shown price swings between $2 and $14 per thousand cubic feet of gas during the past decade. In contrast, relatively more modest price swings for gasoline between $2 and $4 per gallon caused widespread outrage among the commuting public and threatened the collapse of the U.S. automotive industry. As long as natural gas continues to be important for heating homes and businesses, this volatility is likely to continue.

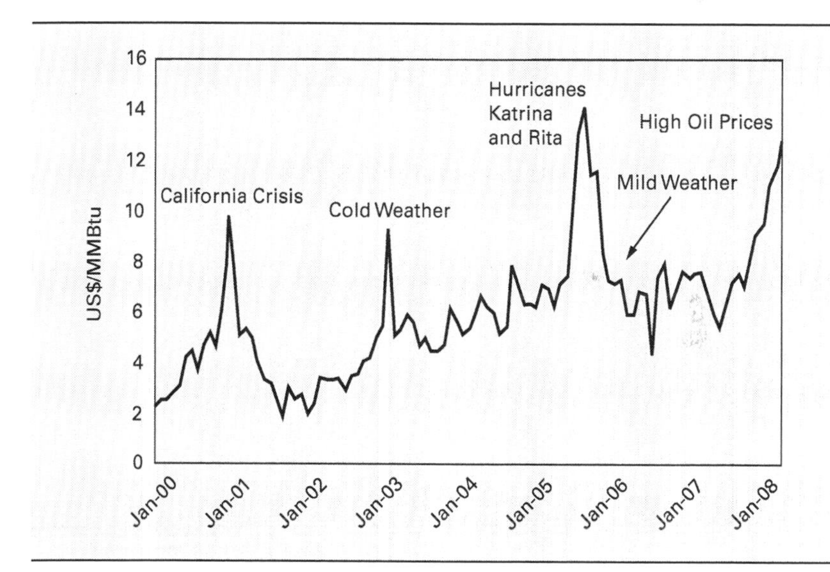

What was T. Boone Pickens thinking? His plan to divert natural gas for use as transportation fuel would exacerbate "gas price" volatility. In the first decade of the twenty-first century, natural gas experienced price variations as large as 600% while gasoline varied a relatively modest 100% over the same time period. *Source: Reference 8.*

Coal

Although the United States may be a bit player when it comes to petroleum reserves, it is the Saudi Arabia of coal. Twenty-seven percent of world coal resources are located in the United States. The combined coal resources of its two closest competitors, China and Russia, barely exceed the U.S. coal reserves. The coal industry likes to point out that at current rates of consumption, U.S. coal reserves would last 250 years. If we were to expand the use of coal to substitute for the petroleum imported into the United States, the coal supply would be reduced to 115 years. This is still adequate for several generations of Americans.

But who wants to run their automobile on coal? Although Amédée-Ernest Bollée built a "coal car" as early as 1875, it wasn't long before liquids were determined to be better automotive fuels. Even in the nineteenth century coal was considered a notoriously dirty fuel, emitting soot, tar, and sulfurous compounds from the steam engines fired on coal. Exacerbating these problems was the inconvenience of shoveling fuel into a boiler. For these reasons, the Stanley Steamer that followed Bollée's coal-fired contraption used cleaner burning kerosene or gasoline to fire the boilers in these steam-powered automobiles. If the world is to use its plentiful supplies of coal as alternative transportation fuel, we will not only have to learn how to convert it into a suitable energy carrier like hydrogen, Fischer-Tropsch liquids, or electricity, we will have to figure out how to scrub out pollutants like sulfur, nitrogen, mercury, and particulates from the effluent of the manufacturing plant. Removing these pollutants is relatively easy compared to recovery and disposal of carbon dioxide released when coal is burned or otherwise processed.

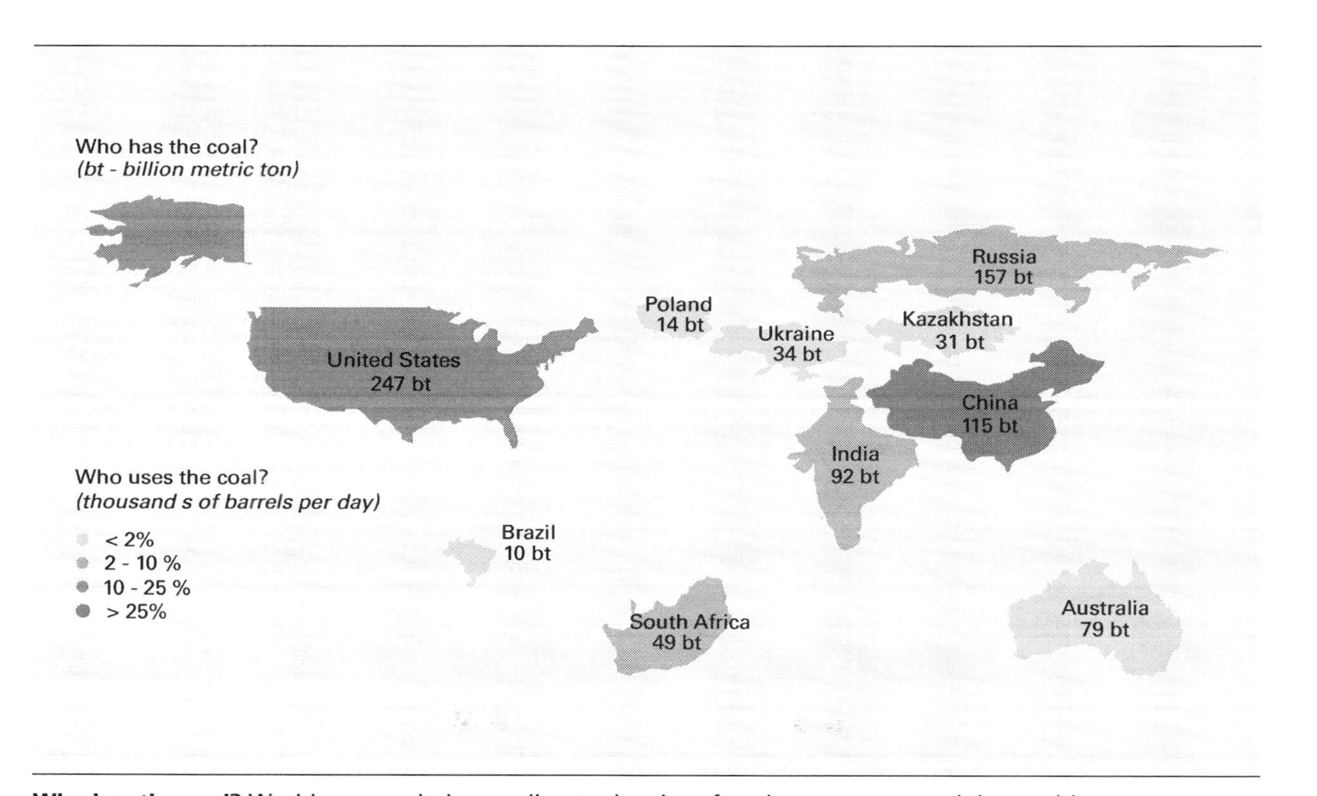

Who has the coal? World map scaled according to the size of coal reserves around the world.

Source: Prepared from data in the 2007 BP Statistical Review of World Energy.

When fossil fuels are burned they release carbon dioxide. The amount depends upon the distribution of carbon-carbon and carbon-hydrogen bonds in the fuel molecules. Coal, with a relatively larger fraction of carbon-carbon bonds, emits 15% to 40% more carbon dioxide per unit of energy than petroleum when these fuels are burned for the generation of electricity. The problem is exacerbated if coal is used to produce synthetic gasoline and diesel fuel. In a process known as coal-to-liquids (CTL), about 2.0 gigajoules of energy in the form of coal are required to produce one gigajoule of energy in the form of synthetic gasoline or diesel. In comparison, petroleum refining consumes only 1.4

Jumping the tracks: Although Amédée-Ernest Bollée demonstrated his "coal car" in 1875, coal was soon relegated to rail transportation and stationary power. *Source: Drawn from a public domain photograph.*

gigajoules of energy in the form of petroleum to produce one gigajoule of gasoline or diesel fuel. Accounting for the energy to extract, transport, and process the coal can yield life-cycle emissions that are at least 100% higher than for gasoline from petroleum. It would be unfortunate if reducing dependence upon imported petroleum resulted in higher emissions of greenhouse gases.

Of course, carbon dioxide generated at a CTL plant could be captured and sequestered to dramatically reduce greenhouse gas emissions. Carbon capture and sequestration (CCS), however, only recovers about half of the carbon from the processed coal with the remaining carbon incorporated into the liquid fuel as hydrocarbons. This carbon is released to the atmosphere as carbon dioxide when the hydrocarbons are burned in an automobile engine. Depending upon the details of the process, the life-cycle greenhouse gas emissions for CTL with CCS may only be marginally better or even worse than those for gasoline from petroleum. A preliminary report by the U.S. EPA suggested that even with CCS, the life-cycle greenhouse gas emissions for CTL was 4% worse than gasoline from petroleum.[10] Based on this finding, lawmakers added a provision to the Energy Independence & Security Act (EISA) of 2007 that precludes United States government agencies from buying oil produced by CTL and other processes that generate more greenhouse gas emissions than petroleum-derived fuels.[11] Early in 2009, the U.S. DOE National Energy Technology Laboratory released a report indicating that CTL with CCS was 5% better than gasoline from petroleum and that application of more aggressive CCS could achieve a 12% improvement compared to gasoline.[12] Of course, these marginal improvements in greenhouse gas emission come at considerable cost. The DOE study indicated that CTL with CCS is not economical until petroleum is $80 to $100 per barrel.

Coal faces tremendous political challenges. As Stephen Chu, Secretary of Energy under the Obama administration,

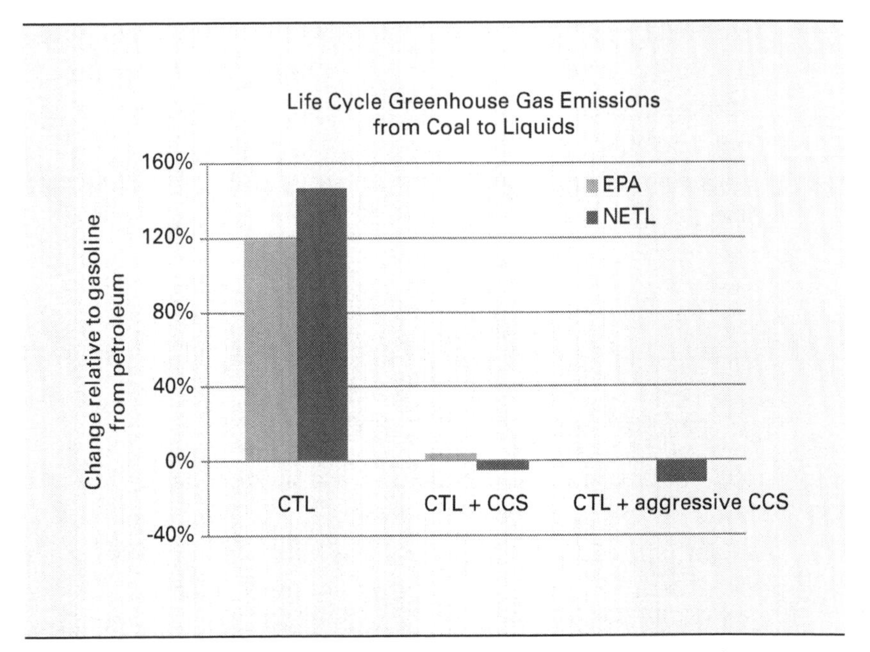

Coal to liquids (CTL) technology will require carbon capture and sequestration (CCS) to achieve life-cycle greenhouse gas emissions that are at best marginally better than gasoline from petroleum. *Sources: References 10 and 12.*

recently confessed, "Coal is my worst nightmare."[13] He is not optimistic that storing billions of tons of carbon dioxide underground or in the ocean depths is economically feasible. "It's not guaranteed we have a solution for coal," he has publicly stated. Gaseous carbon dioxide will take up considerable space even if stored under pressure. It will have a natural tendency to percolate or bubble up from wherever it is stored. That possibility is certain to bring legal challenges to any efforts to launch large-scale carbon dioxide sequestration. As further described by Secretary Chu,[14] "Because there would be people saying I don't want this done in my back yard because if the carbon dioxide ever does bubble to the surface, it could actually kill people." The threat is real. Although carbon dioxide is not toxic in the sense of being poisonous, in sufficient concentrations it can cause suffocation. This was illustrated tragically in Cameron on August 21, 1986 as reported by BBC News:[15]

> At least 1,200 people are feared dead in Cameroon, West Africa, after a cloud of lethal gas escaped from a volcanic lake. The tragedy happened at Lake Nyos, about 200 miles (322 km) northwest of the capital, Yaoundé, during the night. Most of the victims died in their sleep. The gas killed all living things within a 15-mile (25km) radius of the lake, and the area is still highly contaminated. Eventually the cause of this disaster was determined to be carbon dioxide degassing from the bottom of the lake, where it had accumulated by natural causes. Over 1,700 people and 3,500 livestock suffocated.

Unconventional fossil fuels

Tar sands

Tar sands (also known as oil sands) are deposits of sand or clay containing mixtures of water and heavy, viscous polycyclic aromatic hydrocarbons known as bitumen. Oil sands may represent as much as two-thirds of the world's total petroleum resource.[16] Major deposits exist in the Athabasca Oil Sands in Alberta, Canada (1.7 trillion barrels) and the Venezuelan Orinoco oil sands (1.2 trillion barrels). The dif-

When nature has gas: Lake Nyos in Cameroon, West Africa where a natural release of carbon dioxide suffocated 1,700 people in 1986. *Source: Drawn from a photograph of Lake Nyos.*

ficulties of extracting this unconventional fossil resource have discouraged its exploitation although rising petroleum prices have increased interest in its commercial potential. Fully 44% of Canada's oil production in 2007 came from tar sands. The high viscosity of bitumen requires tar sands to be extracted by strip mining or by injection of steam or solvents in the deposits to reduce their viscosity. Recovery and processing requires more water and energy than conventional oil extraction, which increases its environmental impact. For example, life cycle greenhouse gas emissions for production and use of gasoline from tar sands are as much as 40% higher than for petroleum-derived gasoline.[17]

Oil shale

Oil shale is a fine-grained sedimentary rock containing a viscous mixture of hydrocarbons known as kerogen.[18] Unlike bitumen, it is not soluble in normal organic solvents and must be recovered from the shale by heating. This is accomplished by either strip-mining or underground mining the shale and heating it in above-ground retorts to release the kerogen or heating the bedrock to release the kerogen in a process known as in-situ retorting. In both instances the product of retorting must be further processed before it can be sent to a refinery for production of gasoline and diesel fuel. Deposits of oil shale are widely distributed around the world, including major deposits in the United States. Estimates of global deposits range from 2.8 to 3.3 trillion barrels of recoverable oil.[19] Deposits in the United States constitute 62% of world resources. Deposits in the United States, Russia and Brazil account for 86% of the world's shale-oil resources.

Above-ground retorting has significant environmental costs including disturbance of the mined land, disposal of spent shale, use of water resources, disposal of wastewater, sulfur emissions during retorting, and increased greenhouse gas emissions compared to traditional petroleum refining. Some oil shale contains carbonate rock, which decompose to carbon dioxide when heated. This could substantially increase the greenhouse gas emissions than otherwise expected. The process is not considered economical unless the price of petroleum is higher than $60 per barrel as the increased expenses behind the extraction and processing result in crude that is estimated to cost $20-30 more per barrel to produce than that extracted through drilling.[20] The life-cycle greenhouse gas emissions are 36 to 68% higher than those from conventionally produced petroleum-based fuels.[21]

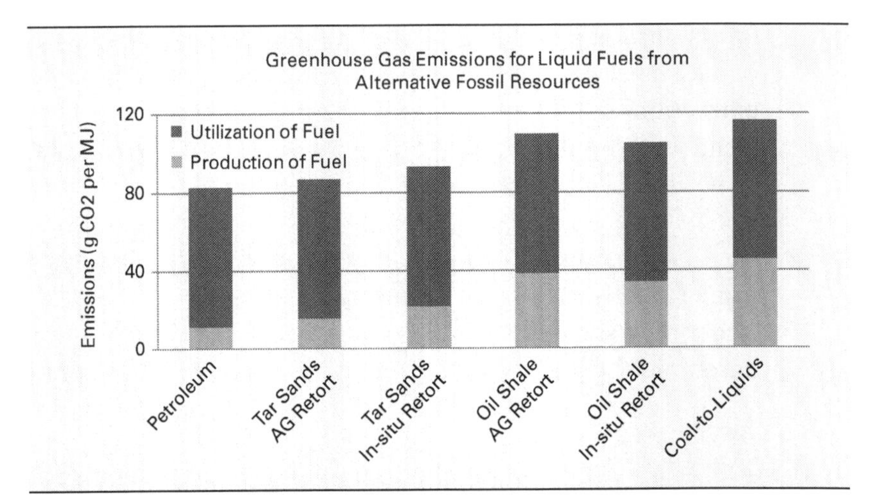

Greenhouse gas emissions associated with production of fuels from alternative fossil resources can be significantly higher than for conventional petroleum resources. The overall greenhouse gas emissions for production and utilization are only marginally higher since utilization of the fuel in vehicles is responsible for most of the greenhouse gas emissions (AG – above ground, In-situ – below ground). *Source: Based on References 17, 21, and 22.*

In-situ processing, which uses electrically-heated deep vertical boreholes in the oil shale deposits, is still under development by Shell Oil Company. The oil shale is gradually heated over a period of two to three years to 650 to 700°F at which point the kerogen is released from the shale and seeps into collection wells in the rock where it can be pumped to the surface. Although the process is energy intensive, Shell estimates that 3.5 units of energy in the form of synthetic crude oil are recovered for every unit of energy required to recover the product. This represents an overall energy efficiency of 78%, which is comparable to recovery of heavy crude using steam injection to enhance oil recovery, and more efficient than strip mining and above-ground retorting of oil shale, which is 63 to 74% efficient. In comparison, extraction of conventional petroleum is at least 80% efficient. In-situ processing has the advantages of leaving the shale in place, which reduces land disturbance and water usage and eliminates disposal of spent shale. There are concerns, however, that it may cause groundwater contamination. The life-cycle greenhouse gas emissions are 21%-47% larger than those from conventionally produced petroleum-based fuels.[22]

Methane hydrates

Methane hydrates, also known as methane clathrates or methane ice, is a type of unconventional natural gas trapped in a lattice of water ice.[23] Resembling slush snow, they were first discovered in the permafrost of the Arctic. Methane hydrates have since been found in shallow marine environments and are thought to be widely distributed around the world. The maximum amount of methane stored in the water lattice structure is 1 methane molecule per 5.75 molecules of water. One liter of methane hydrate solid would release about 164 liters of methane gas.[24] Methane hydrates are thought to have formed from anaerobic decomposition of organic matter by bacteria. The released gas migrated along geological faults where it eventually reacted with cold sea water to form hydrates. Early estimates of methane hydrate reserves were as much as a ten thousand times greater than all other global reserves of natural gas.[25] Improved understanding of how methane hydrates are formed have decreased these estimates to 35 to 175 quadrillion cubic feet of gas,[26] which is still 200 to 1000 times greater than other global reserves of natural gas. Recovery represents distinct challenges to their commercial exploitation. Most methane hydrates are found at low concentrations (0.9-1.5% by volume) and technology for mining them has yet to be developed. Unlike conventional natural gas deposits, which are trapped in impermeable reservoirs, methane hydrates are trapped by virtue of the temperature and pressure at which they exist. If not done carefully, attempts to recover methane from hydrates could destabilize the deposits, resulting in massive accidental releases of methane to the atmosphere. Because methane's global warming potential is 21 times greater than carbon dioxide, these unintended emissions could be a potent contributor to global climate change.

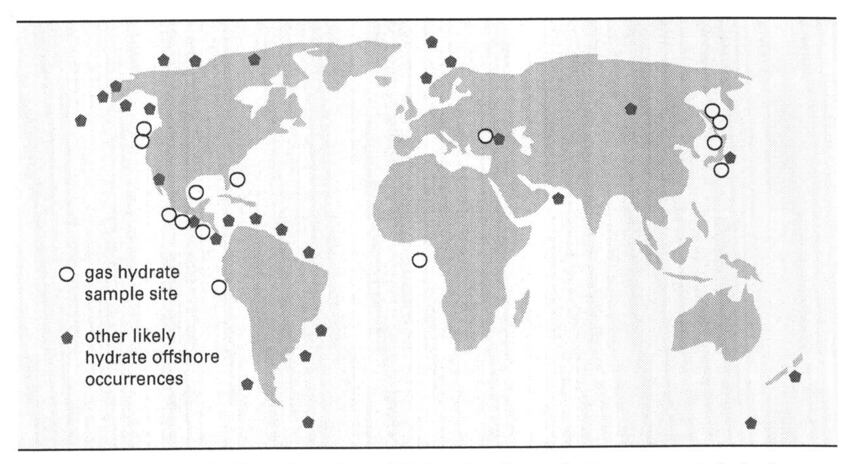

Gas hydrates are believed to be widely distributed along coastal shelves around the world. *Source: Reference 25.*

Nuclear power

Considering the safety concerns with nuclear energy, especially in light of the devastating Fukushima reactor meltdown in 2011, no one today would think of nuclear energy as an "energy carrier" for automotive transportation. And yet in the 1950's and 1960's, while the aerospace industry was experimenting with nuclear-powered airplanes, Ford Motor Company introduced the Nucleon, a concept car with a fission reactor in the trunk. But like coal, fissile material is better left back at the central energy station where electricity or hydrogen can be securely generated as energy carriers.[27]

Fissile material in your tank: Ford's Nucleon concept car was powered by a nuclear reactor. *Source: Drawn from Ford Motor Company publicity photograph.*

The size of nuclear resource depends upon the evolution of technology and policy in the coming years.[28] Non-proliferation policies in the United States artificially limit the supply of fissile material to the relatively rare uranium-235, which accounts for only 0.72% of natural uranium. At current rates of consumption, there is enough uranium-235 to supply power plants based on light-water reactors for one hundred years. A good safety record was marred by the meltdown at the light-water reactor at Fukushima, Japan in 2011. Furthermore, they are relatively inefficient in the use of uranium resources and generate highly radioactive transuranic elements in the spent nuclear fuel, which must be isolated for hundreds or even thousands of years.

The use of so-called breeder reactors would vastly extend the supply of fissile material by converting uranium-238, the balance of natural uranium, into plutonium-239, a fissile material suitable for either fueling nuclear reactors or constructing nuclear bombs.[29] The uranium supply becomes essentially inexhaustible through the use of breeder reactors, if the depleted fuel is reprocessed to recover the plutonium. The breeder reactor cycle also "burns up" more of the transuranic material, reducing the amount of long-lived nuclear wastes that must be stored. Despite these advantages, Presidents Gerald Ford and Jimmy Carter decided to abandon the U.S. breeder reactor program and nuclear fuel reprocessing as part of nuclear non-proliferation measures.[30] The concern was that plutonium from the reprocessing of spent nuclear fuel would find its way into the nuclear weapons. In 2001, President George W. Bush's National Energy Policy recommended resumption of efforts to develop reprocessing and fuel treatment technologies that are proliferation-resistant. Although efforts are underway to modify fuel reprocessing, currently all spent nuclear fuel is treated as waste.

While emerging nations like China and India plan on using nuclear power to help expand their rapidly growing economies, the future of nuclear power in the United States is still in doubt. The close association of nuclear technologies for power and weapons has tarnished the image of the nuclear industry. Fear of nuclear accidents and nuclear

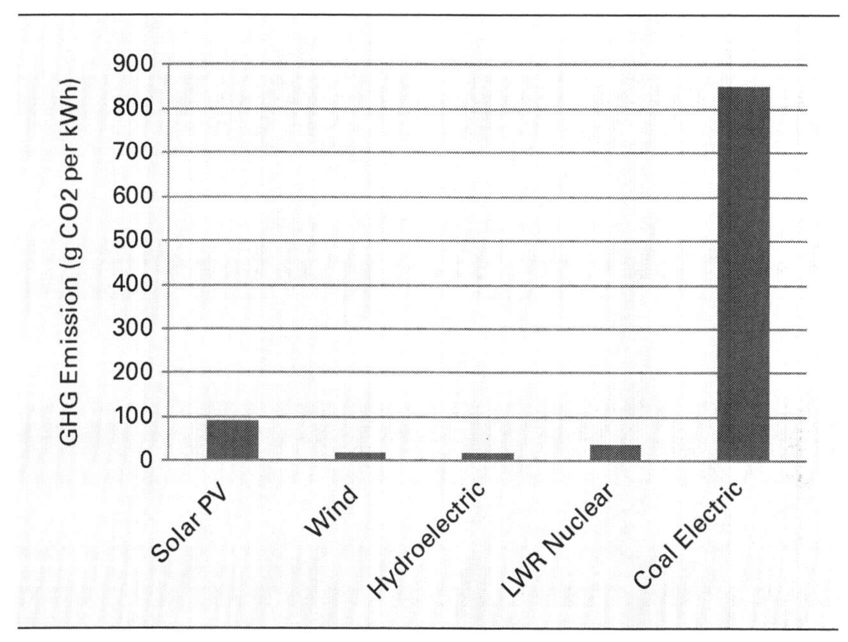

Nuclear power has life cycle greenhouse gas emissions that are comparable or even lower than renewable electricity (GHG – greenhouse gases; kWh – kilowatt hour; PV – photovoltaic; LWR – light water reactor). *Source: Reference 31.*

proliferation and competition from inexpensive fossil fuels conspired to cripple the nuclear power industry. Ironically, the United States is the largest producer of nuclear power in the world although growth of this industry has been essentially flat for several years. Concerns about global climate change may change its pariah status, although a recent focus on safety concerns has caused some developed countries, such as Germany and Japan, to reduce their dependence on nuclear power. Because nuclear fission does not "burn" fuel in the traditional sense of combusting it, no carbon dioxide is emitted from a nuclear power plant. Of course, the construction of a nuclear power plant and the production of fissile materials consume fossil fuels and these must be accounted for in a life-cycle assessment of the greenhouse gas emissions related to nuclear power generation (a similar accounting applies for renewable resources like solar photovoltaics, wind, and hydropower). This kind of analysis reveals that greenhouse gas emissions for nuclear power are almost ten times lower than for coal-fired electric power.[31]

It remains to be seen whether the technical and political barriers to wider use of nuclear energy can be overcome. Meanwhile, high-level radioactive wastes accumulate in 121 temporary, above-ground storage locations scattered across the United States, waiting for a permanent solution to our lack of either permanent storage or reprocessing facilities. One hundred sixty-one million Americans reside within 75 miles of these sites.

Renewable energy

Renewable energy is any perpetual (that is to say, lasting for an indefinitely long period of time) energy flow that can be harnessed for useful purposes. With the exception of geothermal and tidal sources, all renewable energy is a form of solar energy, either direct, which includes solar power and photosynthesis, or indirect, including wind, hydropower, and wave power. Geothermal energy is heat energy generated within the Earth by gravitational and natural radioactive processes. Tidal energy is water-in-motion arising from the gravitational interaction of the Earth and Moon and is distinct from wave energy.

Many people have the misperception that the resource base for renewable energy is small. In fact, the annual flow of renewable energy within the boundaries of the United

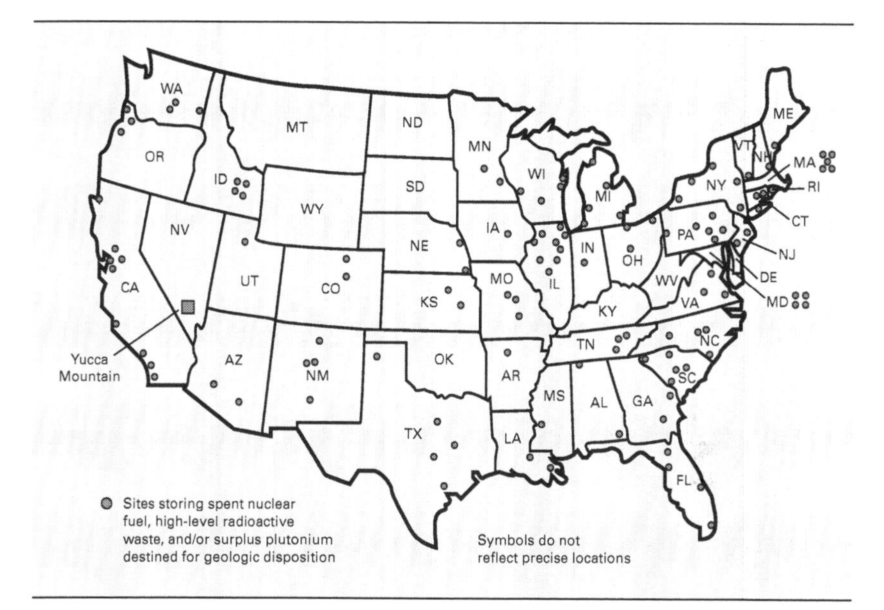

Not in my backyard: High-level radioactive wastes are scattered across 121 locations in the United States, waiting for a better approach to nuclear energy policy. The Obama administration has recently decided not to use the Yucca Mountain permanent disposal site in Nevada. *Source: U.S. Department of Energy.*

States overwhelms the amount of fossil resources buried in its geological deposits. The U.S. Department of Energy estimates that the total fossil energy resource within the United States is 250,000 quads,[32] which is enough primary energy to last us about 250 years at the current rates of consumption (quads are frequently used to measure U.S. national energy production and consumption in the United States; it is one quadrillion British thermal units or 1.055 exajoules of energy). The annual flow of renewable energy includes geothermal energy, direct solar energy (including biomass), and wind energy. Within the United States the renewable energy resource is 3.6 million quads per year, or 36,000 times the amount of primary energy we consume every year. This disparity between renewable and fossil energy resources is substantially reduced when considering the practical difficulties of harnessing renewable energy flows. The amount

Total United States Primary Energy Resources.*

	Total resource (quads)	Technically recoverable (quads)	Economically recoverable (quads)	Annual production (quads)
Renewable resources				
Geothermal	1,497,925	22,782	247	0.35
Direct solar	1,034,940	586,687	352	3.66
Wind	1,026,078	5,046	5	0.32
Hydroelectric	986	157	58	2.46
Fossil resources				
Shale oil	159.604	11,704	1	0.00
Coal	87,458	38,147	5,266	23.48
Petroleum	2,767	1,102	156	10.80
Natural gas	1,705	887	231	22.22
Peat	1,415	354	–	0.00
Nuclear resources	1,177	731	42	8.41

Source: DOE EIA and BP

*Renewable resources are energy flows that are annually available; fossil and nuclear resources are total geological accumulations of energy resources. Total resource – includes resources that are not technically recoverable today; technically recoverable – includes resources that may not be economically recoverable today; commercially recoverable – technically and economically recoverable; annual production – actual production numbers from BP Statistical Review of World Energy.[33]

of renewable energy technically recoverable is about 600 times our annual primary energy consumption, but most of this could not be recovered cost-effectively. The amount of renewable energy that could be economically recovered is estimated to be about six times our annual primary energy consumption. New technologies could enhance the technical and commercial prospects of renewable energy but clearly it is more than sufficient for our current energy needs. The reality is that we utilize very little of this renewable energy resource. In the United States it represents only 10% of the annual production of primary energy whereas it could supply 100% of our needs.[33]

One of the great challenges of renewable energy resources is that they are diffuse forms of energy. We have become accustomed to calling upon concentrated geological deposits of fossil fuels for our energy needs. Twentieth century civilization was unique in the history of human society in its reliance on energy sources that deliver high energy fluxes upon demand. Energy flux is the rate energy is delivered to or from a unit of land area, usually measured in units of watts per square meter. This metric recognizes that land resources, whether derived from fossil or renewable resources, must be devoted to energy production. Coal-fired electricity requires land for coal mines, power stations, and transmission lines. Residential solar photovoltaic power systems require roof-tops to mount solar collectors. Clearly, energy systems with high energy fluxes have smaller land requirements and can more easily supply densely populated cities or energy intensive industries.

Fossil resources have the advantage in terms of energy fluxes. Geological processes operating over eons of time concentrated fossil resources as chemical energy in the form of coal, oil, and natural gas with energy densities in the range of 1,000 watts per square meter (W/m^2) to 10,000 W/m^2. In contrast, renewable energy fluxes are less than 1,500 W/m^2 and typically deliver useful energy in the range of 0.5 to 30 W/m^2.[34] As further perspective, consider that energy demand ranges from about 45 W/m^2 for residential applications to 500 W/m^2 for heavy industry uses like steel making. The high energy civilization of the twentieth century diffused concentrated fossil energy resources to match the more modest energy flux requirements of its end uses. Future societies built upon renewable energy will have to do the very opposite and concentrate its diffuse energy sources to meet its energy flux requirements. As observed by Vaclav Smil:[34]

> Mismatch between the low power densities of renewable energy flows and relatively high power densities of modern final energy uses means that any large-scale diffusion of solar energy conversions will require a profound spatial restructuring, engendering major environmental and socioeconomic impacts. Most notably, there would be vastly increased fixed land requirements for primary conversions and some of these new arrangements would necessitate more extensive transmission rights of ways. Loss of location flexibility for electricity generating plants converting direct or indirect solar flow and inevitable land-use conflicts with food production would be additional drawbacks.

Although the issue of land use change has recently emerged as a concern about expanded biofuel agriculture, it is a problem facing all renewable energy which we obtain, with the exception of geothermal, from a thin boundary layer at the surface of the Earth known as the biosphere. In some respects, renewable energy will be more intrusive on the landscape than fossil energy. A coal-fired power plant is a point-source of pollution emissions, which can spread across the landscape with local, regional, and global impacts. However, as a point-source, its emissions can be more easily managed and regulated. A bioenergy plantation or solar power plant is spread across the landscape and essentially becomes part of a stable ecosystem. Traditional notions of environmental impact will have to be rethought because renewable energy, drawing upon energy flows in the biosphere, will impact the environment. The challenge is whether this impact will be positive or negative in sustaining natural ecosystems and human societies.

The second challenge of renewable energy is that much of it is intermittent. Since energy demand by consumers and industry varies from day-to-day and season-to-season, we must be able to supply power in a predictable and controllable manner. This is relatively easy to achieve with fossil fuels because this subterranean energy supply is relatively fixed in space and time except when disturbed by human activity. On the other hand, renewable energy is by definition energy flowing through the biosphere, which is subject to variations in space and time. The amount of radiation reaching the Earth from the Sun is essentially constant over thousands of years, but the amount of sunlight reaching the Earth's surface varies both predictably due to the rotation and precession of the Earth and unpredictably due to cloud cover. The amount of indirect solar energy in the form of wind and waves shows significant daily and seasonal variations that are not compatible with generating base load power. Among the major direct and indirect forms of solar energy only sunlight that is photosynthetically converted by nature into chemical energy known as biomass can be harvested and stored in a relatively predictable manner. Rain water stored in glaciers and large reservoirs for hydropower and tidal water trapped in bays by dams are examples of gravitational storage of renewable energy, but the amount of energy that can be stored in this way is relatively small compared to the future energy needs of society.

Although we have today technologies to convert sunlight and wind into power, we have no practical way to store it in the quantities required to achieve a level supply of energy

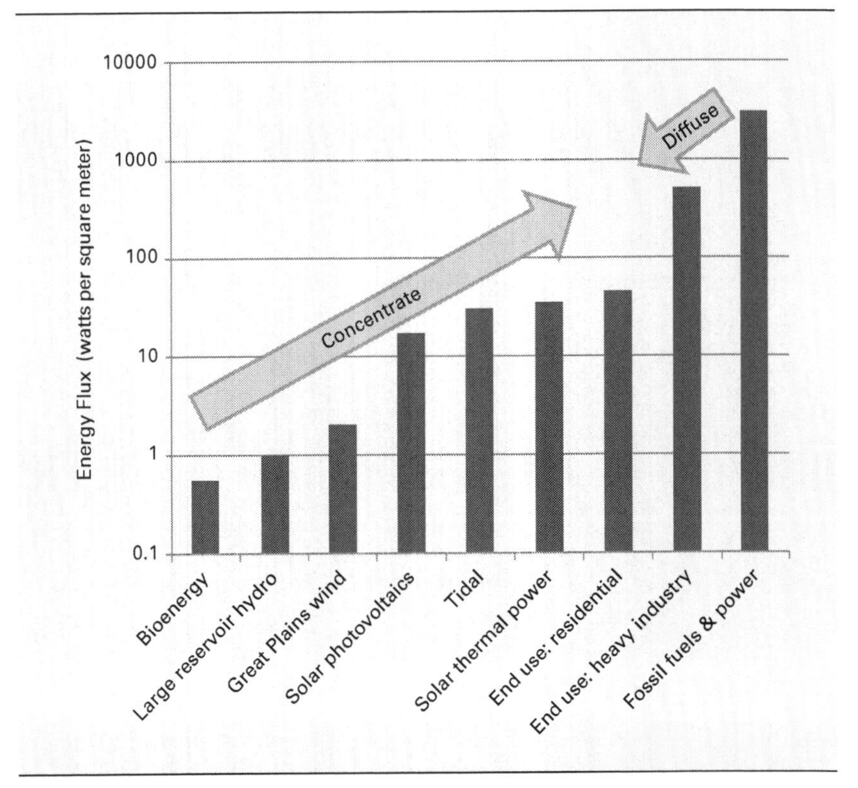

A shift to renewable energy will entail a restructuring of society in order to use these diffuse forms of energy. Notice the logarithmic scale for energy flux. *Source: Based on Smil.*[34]

without building generating capacity many times greater than peak power demand. Other than biomass, the two most promising approaches to energy storage are electric batteries and hydrogen. Neither electricity nor hydrogen is a primary energy source. Their prospects as energy carriers are described in Chapter 5.

5

What are our alternatives to gasoline?

Energy carriers

Energy carriers are convenient and "clean-burning" forms of energy that are manufactured from primary energy sources. One common misperception about the manufacture of energy carriers, which is often heard in discussions about the "energy balance" of biofuels, is that one should expect to get more energy out of fuels than goes into their manufacture. In fact, one should expect just the opposite. For all real energy conversion processes some energy is dissipated as heat or diverted to co-products other than the desired energy products with the result that the useful energy out is less than the total energy into the process. This is the basis for the engineering definition of energy efficiency on a percentage basis:

Energy efficiency (%) =
(useful energy out/total energy into the process) x 100

That the useful energy obtained from a process is always smaller than the energy expended is a consequence of the First and Second Laws of Thermodynamics. By way of analogy, a dropped rubber ball may be expected to recover some of its original elevation (gravitational potential energy) on the rebound, but no one seriously expects it to bounce higher than the height from which it was dropped. Similarly, to meet the often expressed sentiment that "one should get more energy out of biofuels than one puts into it" would mean the energy efficiency of the process was greater than 100%, which is impossible even for petroleum-based fuels.

What matters is preserving within an energy carrier as much as possible of the primary energy used to manufacture it, subject to the constraints of the laws of thermodynamics and the cost of accomplishing this purpose. Success in this endeavor is strongly dependent upon the nature of the energy source, the quality of the energy carrier to be produced, and the capital investment that can be economically justified. Refining sweet light crude to gasoline can recover up to 72% of the primary energy source. On the other hand, burning coal in a steam power plant to produce electricity transfers as little as 33% of primary energy source into the energy carrier. Conversion of biomass into biofuels is expected to achieve an energy efficiency between these two processes, estimated to be about 50%.

If a molecule is to serve as an energy carrier, clearly it must be able to undergo an energy releasing (exothermic) reaction within an engine or fuel cell. Conversely, we might expect that an energy absorbing (endothermic) reaction is required to manufacture an energy carrier. For example, the decomposition of water to produce hydrogen (along with oxygen) is an endothermic reaction. The energy to produce this energy carrier comes from a primary energy source; for example, solar energy converted to electricity can then power an electrolysis unit to decompose water into hydrogen and oxygen.

In fact, both endothermic and exothermic reactions can be used to produce energy carriers. Consider a primary

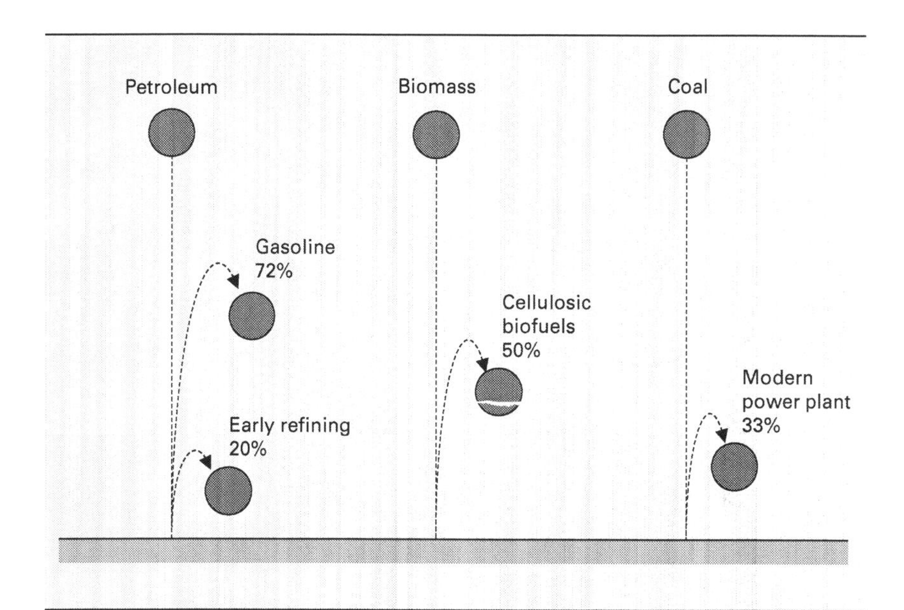

You always get less energy out of a process than you put into it: Rebound of a dropped rubber ball as an analogy for energy efficiency.
Source: Author.

energy source that stores a large amount of chemical energy but in a form not convenient as transportation fuel. An exothermic reaction that rearranges chemical bonds to yield molecules more useful as transportation fuel might be employed despite the loss of part of the original energy. An example is fermentation of glucose to ethanol by yeast, an exothermic process that consumes about 3% of the chemical energy of the sugar to support the metabolism of the microorganism. From the stand point of meeting the energy needs of the microorganism, fermentation is an appallingly inefficient process. However, the rejected energy is in the form of ethanol, an energy dense liquid that is more convenient as transportation fuel than the sugar or starch granules from which it was produced.

Although exothermic reactions are commonly employed to convert energy sources into energy carriers, they are inherently less efficient than endothermic reactions for the production of energy carriers. This is because the energy released during the manufacture of energy carriers is frequently dissipated as waste heat. Nevertheless, many people have the misconception that the addition of energy to support the production of an energy carrier inevitably translates into low energy efficiency, which overlooks the fact that this added energy is incorporated into the energy product. Fuel manufacture is usually a combination of exothermic and endothermic processes, which are ideally balanced to achieve the highest energy efficiency (that is, very little waste energy is rejected from the process).

The rest of the chapter explores different kinds of energy carriers, both those used today and those proposed for the future. A variety of biological and thermal processes can be used to produce them. Later chapters will examine various ways to make these energy carriers from biomass.

Traditional transportation fuels

Traditional transportation fuels are classified as gasoline, diesel fuel, or jet fuel.[1] Gasoline is intended for spark-ignition (Otto cycle) engines; thus, it is relatively volatile but resistant to autoignition during compression. Diesel fuel is intended for use in compression-ignition (Diesel cycle) engines; thus, it is less volatile compared to gasoline and more susceptible to autoignition during compression. Jet fuel is used in gas turbine (Brayton cycle) jet engines, which are not limited by autoignition characteristics but otherwise have very strict fuel specifications for reasons of safety and engine durability. All consist of mixtures of hydrocarbons, compounds of hydrogen and carbon, derived from petroleum. They differ in the kinds and amounts of hydrocarbons in the mixtures, which give the fuels distinctive boiling point ranges.

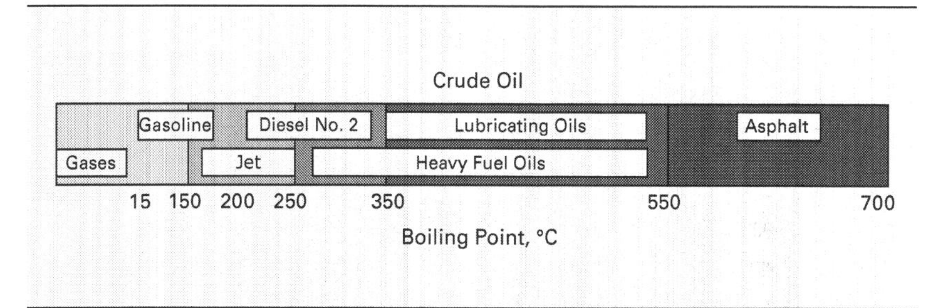

Traditional transportation fuels exploit the different boiling points of hydrocarbons that make up petroleum. Boiling point increases as the molecular weight of hydrocarbons increases. Large hydrocarbon molecules can be cracked to smaller molecules with lower boiling points that are suitable as fuels although this requires the additional expenditure of energy and equipment.

The hydrocarbons making up traditional transportation fuels include chains of single-bonded carbon atoms known as straight-chain alkanes or paraffins; branched-chain alkanes known as isoparaffins; rings of single-bonded carbon atoms known as cycloalkanes (sometimes referred to as naphthenes), and six-member rings of carbon with alternating single and double bonds known as aromatics.

Gasoline is a mixture of hundreds of different hydrocarbons that contain between 5 and 12 carbon atoms with boiling points in the range of 25–225 °C.[2] Most of the mixture consists of alkanes formulated to meet vapor pressure requirements for the fuel. A few percent of

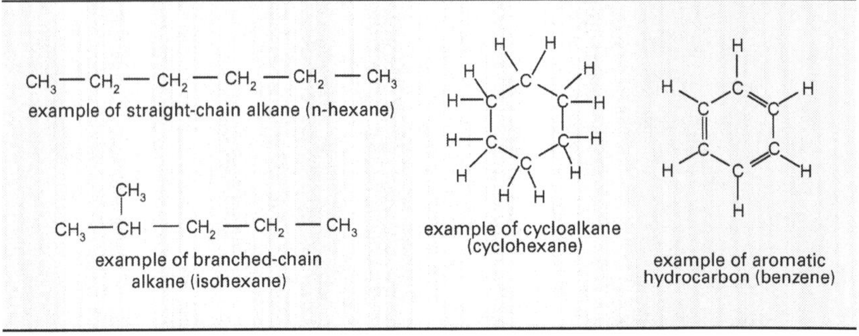

$CH_3—CH_2—CH_2—CH_2—CH_2—CH_3$
example of straight-chain alkane (n-hexane)

$CH_3—CH(CH_3)—CH_2—CH_2—CH_3$
example of branched-chain alkane (isohexane)

example of cycloalkane (cyclohexane)

example of aromatic hydrocarbon (benzene)

Examples of six-carbon atom hydrocarbon molecules found in petroleum-derived transportation fuels. Transportation fuels consist of hydrocarbon molecules containing 5 to 22 carbon atoms arranged in linear chains, branched chains, or rings with a wide range of chemical and physical properties.

aromatic compounds are added to increase octane number, the figure of merit used to indicate the tendency of a fuel to undergo premature detonation within the combustion cylinder of an internal combustion engine. The higher the octane number, the less likely a fuel will detonate until exposed to an ignition source (electrical spark). Premature denotation is responsible for the phenomenon known as engine knock, which reduces fuel economy and can damage an engine. Federal regulation in the United States requires gasoline sold commercially to be rated using an average of the research and motor octane numbers. Gasoline rated as "regular" has a commercial octane number of about 87 while premium grade is 93.

Diesel fuel is a mixture of hydrocarbons containing 12 to 22 carbon atoms with lower volatility and higher viscosity.[2] Because diesel fuel is intended to be ignited by compression rather than by a spark, its autoignition temperature is lower than for gasoline. The combustion behavior of diesel fuels are conveniently rated according to cetane number, an indication of how long it takes a fuel to ignite (ignition delay) after it has been injected under pressure into a diesel engine. A high cetane number indicates short ignition delay; for example, No. 2 diesel fuel has cetane number of 37 – 56 while gasoline has a cetane number less than 15.

Jet fuel is designated as either Kerosene Type (including Jet A, Jet A1, and JP8) or Wide Cut Type (including Jet B and JP4). Kerosene Type is a mixture of straight-chain alkane (paraffin) molecules containing 9 to 20 carbon atoms giving it a relatively high boiling range of 165°-290°C, making it similar to diesel fuel. It also contains a certain percentage of aromatic compounds to meet desired fuel performance characteristics. Wide Cut Type jet fuel is a blend of kerosene and gasoline containing 5 to 15 carbon atoms, which gives it a wider boiling point range. It contains straight-chain and branched-chain alkane molecules as well as aromatic hydrocarbons. Although its lower boiling point range enhances cold-weather performance, it also makes it more flammable and dangerous to handle. At one time it was preferred in the United States because of the wide availability of gasoline for blending with more scarce kerosene. However, safety and environmental concerns have resulted in the phase out of Wide Cut Type blends for both commercial and military aviation except in demanding applications.

Traditional transportation fuels are, by definition, blends of various hydrocarbons obtained from refining petroleum. However, alternative primary energy sources, such as coal or biomass, could also be used to produce hydrocarbon fuels although they likely would consist of different blends of hydrocarbons than are obtained from petroleum. Synthetic hydrocarbons from biomass are subsequently discussed in this and later chapters.

Gaseous transportation fuels

The ideal transportation fuel is a stable liquid at ambient temperature and atmospheric pressure which readily vaporizes and burns within an internal combustion engine. Nevertheless, several gaseous compounds have been put forward as transportation fuels because of their ready availability or attractive economic or environmental benefits. These include hydrogen, synthetic natural gas, ammonia, and dimethyl ether.

Hydrogen was widely touted as the future of transportation during the first term of President George W. Bush, with significant federal resources directed toward the development of technologies to generate, store, and utilize hydrogen. The promise of hydrogen was low pollution emissions and high energy efficiency from fuel cells, which employ relatively low temperature electrochemical reactions to power electric vehicles in place of lower efficiency internal combustion engines. Hydrogen can be generated from greenhouse-gas neutral renewable fuels or even from fossil fuels if a practical method for sequestering the resulting carbon dioxide can be devised. Hydrogen fuel generates no tail-pipe pollution emissions since hydrogen generates only water when it reacts with oxygen. Unlike internal combustion engines, fuel cells have the potential to exceed the theoretical efficiency of heat engines. Whereas the best internal combustion engines are limited to thermodynamic efficiencies of 30-40%, fuel cells have efficiencies as high as 60%.[3] Thus, for a unit of energy, a fuel cell vehicle might travel 50% further than a conventional gasoline-powered automobile. Like natural gas, it can be stored as compressed hydrogen gas (GH_2) or liquefied hydrogen (LH_2).

If produced from biomass, hydrogen has production

efficiency and cost advantages compared to biomass-derived liquid fuels, such as cellulosic ethanol, Fischer-Tropsch liquids, and even methanol.[4] These advantages arise from the fact that conversion of biomass into liquid fuels usually involves many conversion steps and may even involve the production of hydrogen as an intermediate product.

Hydrogen as transportation fuel has several shortcomings. Most strategies for early deployment of a "hydrogen economy" depend upon fossil fuels as the energy source for generating hydrogen. For example, coal or natural gas can be chemically reacted with steam to form hydrogen and carbon dioxide. Alternatively, water could be electrolyzed to hydrogen and oxygen with power provided by the electric grid, which relies mostly on coal for its generation, or more efficient and less polluting natural-gas fired combined cycle power plants. The chemical production of hydrogen from fossil fuels has some of the same energy efficiency and cost advantages as production from biomass. However, a transportation system based on electrically-generated hydrogen using fossil fuels for power generation has an overall energy consumption that is 50% to 100% higher than a gasoline-based transportation system. Fossil-fuel generated hydrogen, whether by chemical or electrical processes, also produces copious greenhouse gas (GHG) emissions, not at the tail-pipe of the vehicle, but at the processing facility where energy is extracted from the fossil fuels.

Such energy and GHG disadvantages become apparent through a "well-to-wheels" (WTW) analysis of a proposed transportation system. A WTW analysis considers energy requirements and GHG emissions associated with both production and utilization of an energy carrier.[5] On the production side, all energy use and greenhouse gas emissions are considered, including those associated with extraction and transport of the primary energy source to the processing facility as well as carbon dioxide emitted in the manufacturing process. Overall energy consumption is reported as megajoules (MJ) of primary energy consumed in manufacture of sufficient energy carrier to travel one kilometer (km). Greenhouse gas emissions are reported as grams of carbon dioxide (g CO_2) to travel one kilometer (km). The charts below illustrate the relatively poor well-to-wheels

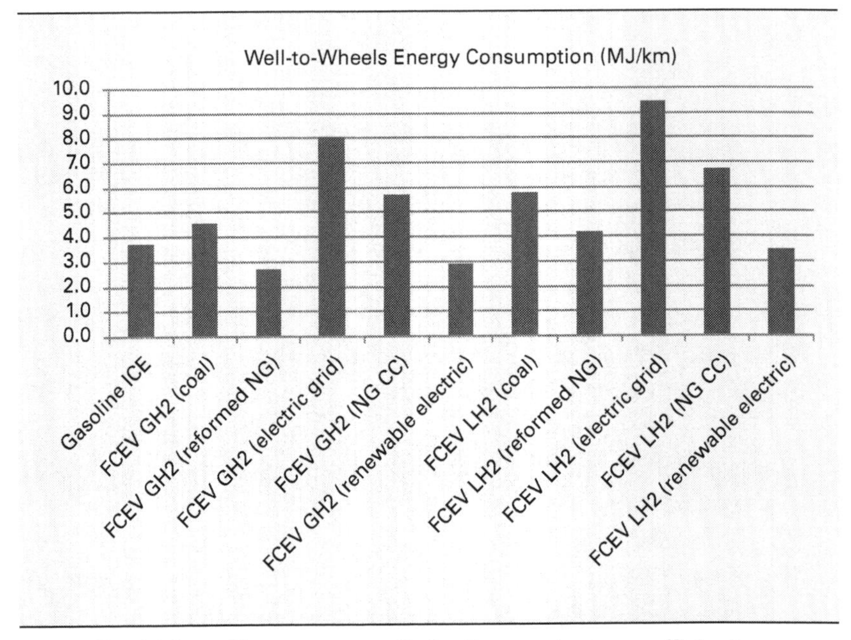

Fuel cell vehicles will not necessarily be the most energy efficient option. Although fuel cells may be more efficient than internal combustion engines, generating hydrogen can be much less efficient than other fuel options. ICE – internal combusiton engine; FCEV – fuel cell electric vehicle; GH_2- compressed gaseous hydrogen; LH_2 – liquefied hydrogen; NG – natural gas; CC – combined cycle. *Source: Graph generated from data in Reference 5.*

energy consumption (MJ/km) and greenhouse gas emissions (g CO_2/km) for fuel cell electric vehicles (FCEV) that use either gaseous hydrogen (GH_2) or cryogenic liquid hydrogen (LH_2) generated from fossil fuels. Certainly carbon sequestration of the carbon dioxide byproduct could dramatically reduce the GHG emissions, although this presents tremendous technical, economic, and political challenges. It would also increase WTW energy consumption.

On the other hand, hydrogen generated from renewable

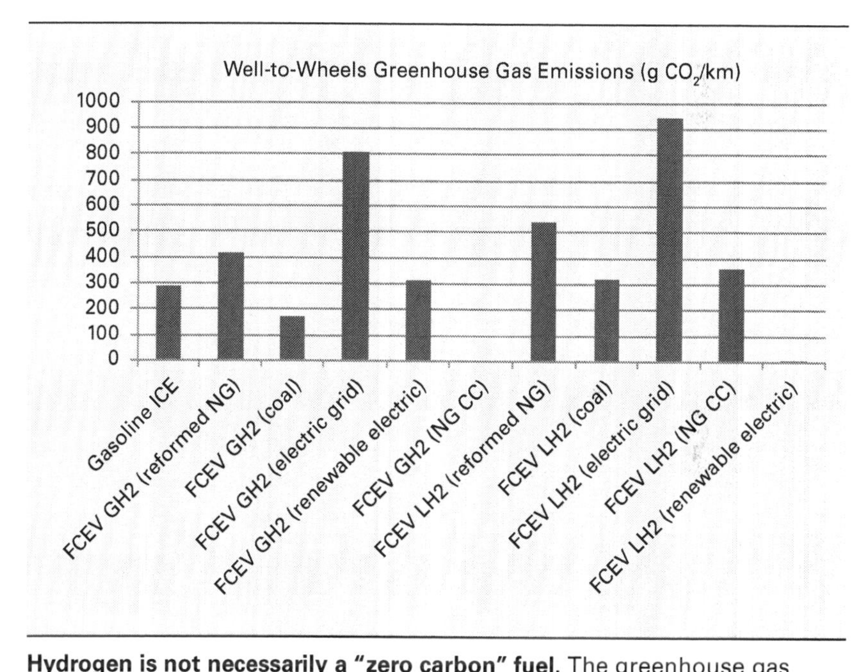

Hydrogen is not necessarily a "zero carbon" fuel. The greenhouse gas profile of hydroge fuel depends upon the primary energy source used to produce it. ICE – internal combusiton engine; FCEV – fuel cell electric vehicle; GH_2- compressed gaseous hydrogen; LH_2 – liquified hydrogen; NG – natural gas; CC – combined cycle. *Source: Graph prepared from data in Reference 5.*

electricity, assumed to be wind power or photovoltaic power, has very low energy consumption and greenhouse gas emissions from a WTW analysis. Although the analysis for the renewable resources neglects the significant energy consumption and GHG emissions associated with the manufacture of these "free energy" harvesting devices, they remain extremely attractive from this perspective. But the cost of renewable hydrogen from wind, sunlight, or geothermal heat is an unattractive \$4 and \$7 per gallon gasoline equivalent.[6]

Even if the cost of generating renewable hydrogen is substantially reduced, there remain the extraordinary challenges of storing and using hydrogen.[7] On a volumetric basis, gaseous hydrogen contains a surprisingly small amount of

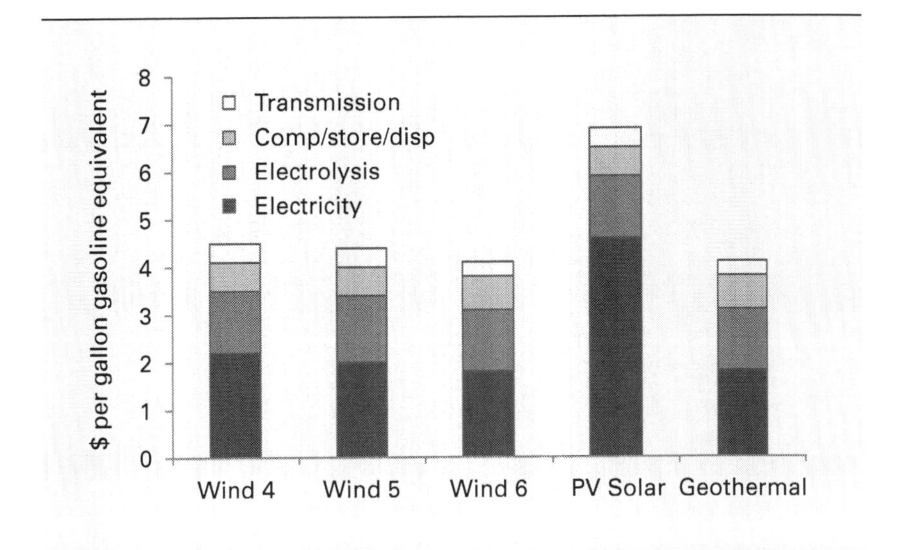

Renewable hydrogen from wind, solar and geothermal energy sources will cost \$4 to \$7 per gallon of gasoline equivalent. Wind Power Classes (50 m elevation): Wind 4 – 500 W/m2; Wind 5 – 600 W/m^2; Wind 6 – 800 W/m^2. PV Solar – Photovoltaic solar. *Source: Adapted from Reference 6.*

hydrogen as a result of its intrinsically low density. Liquefied hydrogen contains 30% less mass of hydrogen than a comparable volume of ethanol. Although a liquid hydrogen-powered fuel cell electric vehicle might have 50% better fuel economy than a gasoline-powered internal combustion engine vehicle, still the hydrogen-powered vehicle would have only half the range of the gasoline-powered vehicle. Whether compressed gas at 340 atmosphere or cryogenic liquid at -253°C, hydrogen presents unique challenges in distributing it to refueling stations and dispensing it to consumers. In addition to the hazards of high pressures and cold temperatures, hydrogen is extremely flammable, generating a hot, invisible flame that is the bane of firefighters.

Synthetic natural gas (SNG) is the name given to

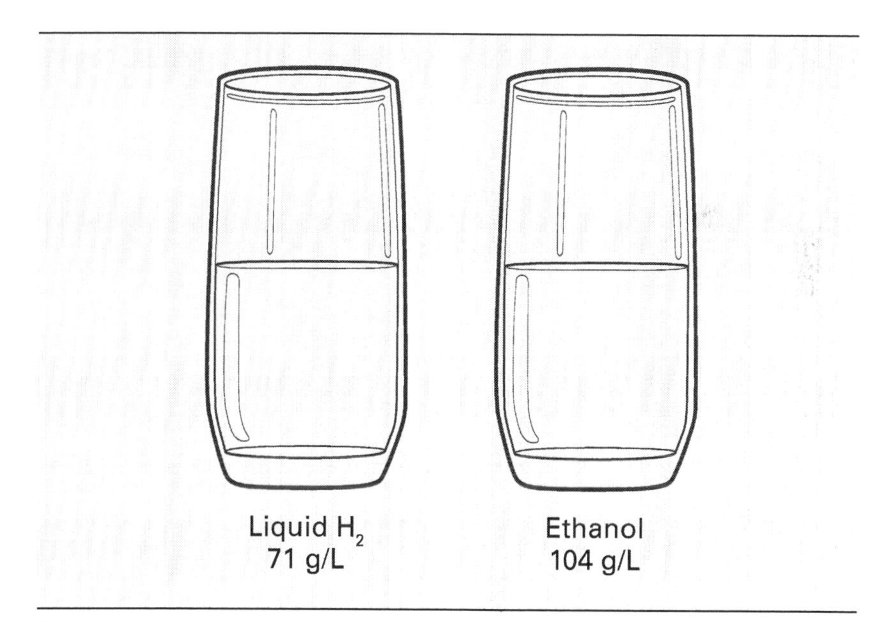

Hypothetical drink of biofuels: A glass of ethanol contains more hydrogen than a glass of liquid hydrogen. *Source: Author.*

methane-rich manufactured gas to contrast it with naturally occurring fossil deposits of methane-rich gas known as natural gas. For all intents and purposes, both are essentially methane, a stable molecule of one carbon-atom bonded to four hydrogen atoms. Methane is both an important fossil energy source (in the form of natural gas) and a potential energy carrier. It has three attractive features as an energy carrier. First, it can employ the natural gas infrastructure already existing in the United States and other parts of the world. Second, methane can be readily generated by a variety of biological and thermal processes using both fossil and biomass resources. Third, it is a clean burning fuel with an excellent safety record, which has encouraged its use in a wide variety of residential, commercial, industrial, and

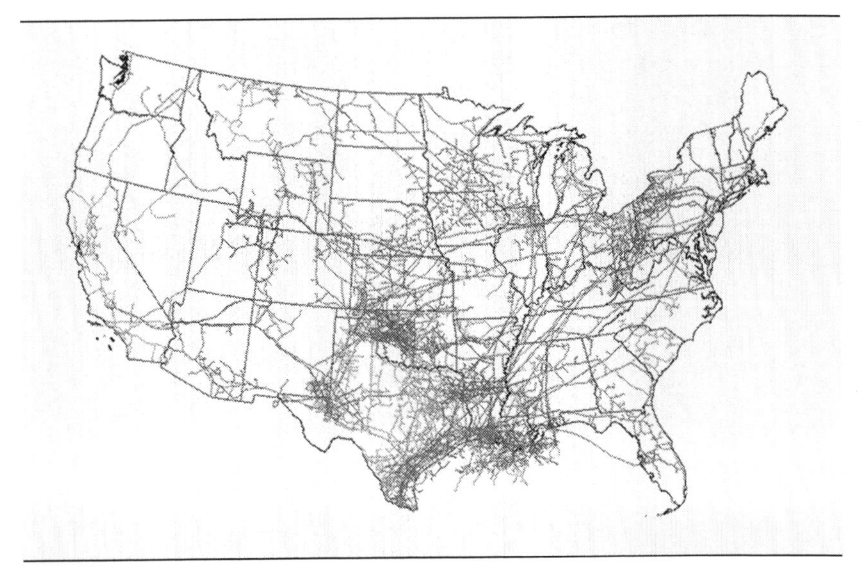

The United States natural gas pipeline network could be used to distribute synthetic natural gas. *Source: Energy Information Administration (Reference 9)*

utility applications. Its primary disadvantage is its low density even under compression compared to liquid fuels. Although the financier T. Boone Pickens has touted compressed natural gas (CNG) as the future of transportation fuels,[8] its low density will likely limit it to urban mass transit. Natural gas can be liquefied (LNG) by compressing and cooling it to cryogenic temperatures, which is an economical method for moving so-called stranded natural gas from remote parts of the world to markets, but LNG is not practical as a motor fuel because of its cryogenic requirements.

Anhydrous ammonia (NH_3) is a gas at ambient conditions but is readily liquefied at room temperature by storage at 10 atmospheres pressure, achieving 87% of the density of gasoline.[10] In fact, it has nearly the same density, boiling point, and octane number as propane, which has been widely employed as a portable fuel supply. The United States already has in place production, storage, and distribution infrastructure for its use as agricultural fertilizer. Ammonia has been tested as fuel in spark ignition engines, diesel engines, and gas turbines. Near theoretical performance was achieved with ammonia by partially dissociating it to obtain 1% hydrogen at the engine intake.[11] Nitrogen oxide emissions were lower than from octane fuel despite the high nitrogen content of ammonia. Despite these advantages, it is far from an ideal transportation fuel. Its low flammability requires the addition of an ignition promoter, which complicates engine design. Because of its lower volumetric heating value, an ammonia fueled vehicle would require a fuel tank about 2.4 times larger than for a propane-fueled vehicle. Ammonia is highly volatile and extremely toxic, further detracting from its use as transportation fuel.

Dimethyl ether (CH_3OCH_3) has been proposed as diesel engine fuel.[12] Dimethyl ether, often identified by the acronym DME, has several attractive features including low production cost, very low soot emissions compared to both traditional diesel and biodiesel fuels, and very high fuel efficiency in terms of the driving range per unit of cropland devoted to biofuels production. Dimethyl ether, like liquefied petroleum gas (LPG) is a non-toxic, flammable gas at ambient conditions that is easily stored as a liquid under modest pressures. It is readily produced from natural gas, coal, or biomass using thermochemical processes described

Gaseous fuels stored as liquids compared to gasoline and ethanol.

	Gasoline	Ethanol	Hydrogen (liq)	Methane (liq)	Ammonia (liq)	Dimethyl ether (liq)
Specific gravity	0.72-0.78	0.794	0.071	0.422	0.682	0.660
Storage pressure (atm)	1	1	5-10	1	6-15	4-7
Storage temperature (C)	ambient	ambient	-253	-163	ambient	ambient
Lower heating value (MJ/kg)	43.5	27	120	49.5	18.8	28.9
Volumetric heating value (MJ/liter)	32.6	21.4	8.5	20.9	12.8	19.1
Special safety concerns	—	—	Highly flammable		Volatile and toxic	—

Source: Various

*Measured at 16C except for liquefied gases, which are saturated liquids at their boiling points.

in Chapter 10. Its main disadvantage is that it would have to be stored under 4–7 atmospheres of pressure and vehicle driving range would be less than most fuels that are liquid at atmospheric pressure. However, it is superior to the other gaseous fuel options because of its combination of low toxicity and ability to be stored as liquid at ambient temperatures.

Alcohols

Alcohols are a group of organic compounds characterized by the presence of one or more hydroxyl groups (OH) attached to the carbon backbone of the molecule. The hydroxyl group is responsible for several prominent physical properties of alcohols including their high solubility in water and high boiling point compared to other compounds of otherwise similar structure. The hydroxyl group is also responsible for alcohols being oxygenated fuels. The alcohols most commonly explored as transportation fuels are methanol, ethanol, and butanol.

Methanol (CH_3OH) is a clear, odorless, flammable liquid suitable as high-octane motor fuel.[13] The fuel properties of methanol include narrow boiling point range, high heat of vaporization, and high octane number. The heating value is only 49% that of gasoline on a volumetric basis. Methanol is very toxic. Recent national legislation banning the closely related and similarly toxic methyl tertiary butyl

$CH_3—OH$ $CH_3—CH_2—OH$ $CH_3—CH_2—CH_2—CH_2—OH$

methanol ethanol butanol

ether (MTBE) as a fuel additive because of concerns about ground water contamination probably has doomed the prospects for methanol as transportation fuel. However, it remains one of the top ten commodity chemicals worldwide and it is a precursor for the production of more promising transportation fuels like dimethyl ether and synthetic gasoline.

Ethanol (C_2H_5OH) has fuel properties similar to methanol although it is considerably less toxic. Also more attractive is its heating value, which is 66% that of gasoline on a volumetric basis. Nevertheless, this suggests a substantial reduction in range for a vehicle that substitutes ethanol for gasoline. In fact, fuel economy depends on several factors affecting the operation of an engine, which some argue improves the performance of ethanol relative to its heating value.[14] For example, the higher octane number for ethanol compared to gasoline (109 vs. 91- 101) allows engines to be designed to run at higher compression ratios, which improves both power and fuel economy. Estimates for efficiency improvements in engines optimized for ethanol instead of gasoline range are 15% to 20%, resulting in a driving range approaching 80% of that of gasoline.[15]

Internal combustion engines can be fueled on pure ethanol (known as neat alcohol or E100) or blends of ethanol and gasoline. Brazil employs 190 proof ethanol (95 vol % alcohol and 5% water), which eliminates the energy-consuming step of producing anhydrous ethanol. In the United States two ethanol-gasoline blends are common: E85 contains 85% ethanol and 15% gasoline and E10 contains 10% ethanol with the balance being gasoline. The advantage of E10 is that it can be used in vehicles with engines designed for gasoline.

Ethanol-gasoline blends have some disadvantages

compared to neat gasoline. Among these are water-induced phase separation and evaporative emissions. Phase separation occurs when water contaminating a storage tank or pipeline is absorbed by ethanol, resulting in a water-rich layer below a hydrocarbon-rich layer, which interferes with proper engine operation. Water contamination is a problem that has not been fully addressed by the refining, blending, and distribution industries; thus transportation of ethanol-gasoline blends in pipelines is not permitted in the United States and long-term storage is to be avoided.[16] Evaporative emissions occur when vapors from volatile fuels are released to the atmosphere during fueling operations. By an unfortunate coincidence, blends of ethanol in gasoline, especially at the low end of ethanol blending (E10 or E15), have higher vapor pressures than either neat ethanol or gasoline. This can be mitigated by sealing fuel storage tanks, but some vapor release is inevitable during fueling operations.

As the carbon chain becomes longer, alcohol becomes less soluble in water and its heating value increases. For this reason, butanol (C_4H_9OH) has some attraction as motor fuel.[17]

Butanol has fuel properties superior to methanol and ethanol and comparable to gasoline.

Fuel	Formula	Energy Content (MJ/L)	Motor Octane Number	Air to Fuel Ratio	Vapor Pressure (psi@100F)
Methanol	CH_3OH	16	104	6.6	4.6
Ethanol	C_2H_5OH	19.6	102	9	2
Butanol	C_4H_9OH	29.2	78	11-12	0.33
Gasoline	hydrocarbons	32	81-99	12-15	4.5

Source: Various

Its energy content is within 10% of gasoline and its octane number, although lower than ethanol, is greater than gasoline's. Its low vapor pressure is attractive in reducing volatile organic matter emissions, but it can make cold starting more difficult. Butanol is only moderately soluble in water, an important advantage compared to fully soluble methanol and ethanol. If butanol fermentation was commercially viable, very likely butanol would be substituted for ethanol as renewable fuel.

Esters

Esters are a common type of plant chemicals, formed by the reaction of organic acid with alcohol. Of particular interest for the production of fuels is the class of esters known as triglycerides, formed by the reaction of a fatty acid with glycerol. A fatty acid is a long-chain organic acid containing an even number of carbon atoms while glycerol is a three-carbon alcohol containing three hydroxyl groups. Triglycerides are more commonly known as fats and oils. The acid fractions of triglycerides can vary in chain length and the number of double bonds present in the chain (degree of saturation).[18]

Compared to petroleum-based diesel fuels, fats and oils have higher viscosity and lower volatility, which results in fouling of engine valves and less favorable combustion performance, especially in direct-injection engines.[19] The solution to this problem is to convert the triglycerides into less viscous and less volatile methyl esters or ethyl esters of the fatty acids, known as biodiesel, and the byproduct glycerol. The fuel properties of biodiesel are very similar to petroleum-based diesel. The specific gravity and viscosity of biodiesel are only slightly higher than for diesel while

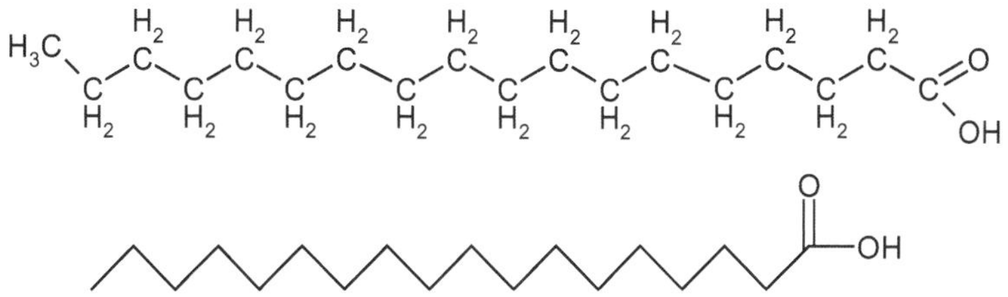

Structural formula and bond-line structure of a fatty acid

Triglyceride (three fatty acid chains attached to a glycerol backbone)

Methyl ester (product of the reaction of fatty acid with methanol)

the cetane numbers and heating values are comparable. Significantly higher flash points for biodiesel represent greater safety in storage and transportation. Biodiesel can be used in unmodified diesel engines with no excess wear or operational problems. Tests in light-duty and heavy-duty trucks showed few differences other than a requirement for more frequent oil changes because of the build-up of ester fuel in engine crankcases. Biodiesel has more cold-performance problems than conventional diesel fuel, which are periodically reported in the press during cold snaps across the upper Midwestern United States. This has been attributed to the presence of minute amounts of naturally-occurring steryl glucosides, which can be mitigated through quality control measures.

Synthetic hydrocarbons

Plant molecules important as feedstock for the production of biofuels include carbohydrates and lipids. As these contain various amounts of oxygen, the focus of most biofuels research and development until recently has been on the production of oxygenated fuels like ethanol and biodiesel (methyl esters). Recognizing the advantages of hydrocarbons as transportation fuels, much current research is directed toward their production from biomass. This includes direct biosynthesis; deoxygenating and cracking carbohydrates, lignin-derived compounds, and lipids; and catalytic synthesis of hydrocarbons from gasified biomass.

Biosynthesis exploits the fact that many plants and microorganisms naturally synthesize hydrocarbons although usually in small quantities. As described in Chapter 9, some researchers are working to promote the metabolic pathways that produce these hydrocarbons with the hope of generating

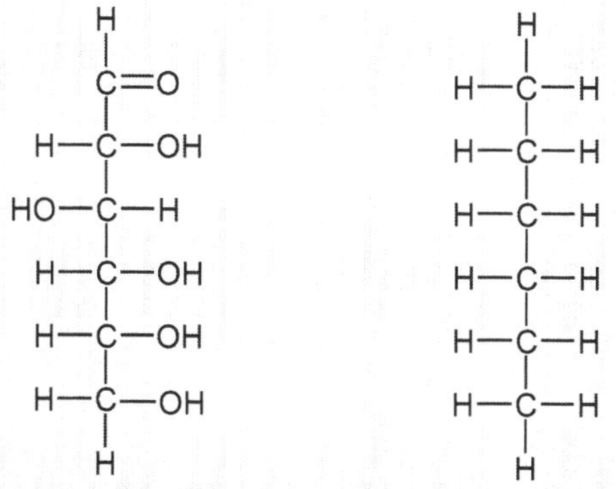

```
      H                    H
      |                    |
      C=O               H—C—H
      |                    |
   H—C—OH               H—C—H
      |                    |
  HO—C—H                H—C—H
      |                    |
   H—C—OH               H—C—H
      |                    |
   H—C—OH               H—C—H
      |                    |
   H—C—OH               H—C—H
      |                    |
      H                    H
```

glucose (a hexose) hexane

Hydrocarbons from carbohydrate: Comparing the molecular structure of a highly oxygenated C6 carbohydrate (glucose) to a highly reduced C6 hydrocarbon (hexane).

commercially significant quantities in either plant tissues or bioreactors.

Deoxygenation is the process of removing oxygen from organic compounds. Remove enough oxygen and an oxygenated organic compound becomes a hydrocarbon. Consider that glucose is a six-carbon sugar (a hexose) with six oxygen atoms attached to it. If hydrogen atoms were substituted for the oxygen atoms along the carbon backbone, the resulting molecule would be a six-carbon alkane known as hexane, one of the lower boiling point constituents of gasoline. The process of reacting hydrogen with a compound to remove oxygen is called hydrodeoxygenation. In addition to hydrocarbons it produces copious quantities of water formed when the hydrogen reacts with the removed oxygen.

Hydrodeoxygenation can be used to strip oxygen from lipids and lignin as well as carbohydrates although the process is expensive in its use of hydrogen: one or two molecules of hydrogen (H_2) are required to replace every atom of oxygen (O). Another approach to deoxygenation sacrifices some of the carbon in the organic compound to remove oxygen. This process, called decarbonylation, removes carbon monoxide (CO) from an organic compound. In certain circumstances, the released CO can react with water to form hydrogen, which can further deoxygenate the organic compound. Deoxygenation is usually promoted with catalysts although purely thermally-driven reactions are possible.

In the case of lipids or lignin, the amount of oxygen to be removed is relatively modest. These biomolecules can be very large with lipids consisting of chains containing 16-18 carbon atoms. Lignin is a true polymer, with chains containing many hundreds of carbon atoms. Recalling that gasoline has chain lengths of only 5 – 12 carbon atoms and diesel fuel has chain lengths between 12 – 22 carbon atoms, some degree of chain-breaking, known as cracking, is required to reduce lipids and lignin to molecular sizes suitable for transportation fuel. This can be accomplished through the action of high temperatures although more desirable results are obtained at more modest temperatures and elevated pressures by the action of hydrogen in the presence of catalysts, a process known as hydrocracking. These processes are more thoroughly described in Chapter 10.

The third approach to producing hydrocarbons from biomass or any carbon-bearing feedstock is to break down complex molecules into their most fundamental building blocks, carbon monoxide and hydrogen, and to synthesize from these energy-rich molecules the desired compounds, which could be either oxygenated compounds or hydrocarbons.

This deconstruction of biomass occurs at elevated temperatures in an endothermic process called pyrolysis. The heat required to drive pyrolysis comes from burning part of the biomass feedstock. This balance between endothermic pyrolysis reactions and exothermic combustion reactions is called gasification. Most although not all of the energy released during combustion is converted to chemical energy in the gaseous product commonly known as syngas or producer gas. Additional products are water and carbon dioxide, so in a sense gasification accomplishes both deoxygenation and decarbonylation without the addition of hydrogen. Chapter 10 describes how carbon monoxide and hydrogen can be reacted over a catalyst to yield hydrocarbons.

Electricity

Electricity is an energy carrier distinct from other energy carriers. Conventional energy carriers are forms of chemical energy whereas electricity is described as "work" by those that study thermodynamics. Unlike chemical energy, there is no theoretical limit to the amount of work that can be transformed into other kinds of energy, which makes it the most valuable form of energy. Other examples of work are the mechanical energy of a rotating shaft and laser light (although not other common forms of light such as sunlight). Electricity is the only form of work that has been widely harnessed as an energy carrier, as evidenced by the electrical transmission grid and battery-powered vehicles and other devices.

Electricity has many advantages as an energy carrier. A well-to-wheels (WTW) analysis of the kind previously described for hydrogen applied to battery electric vehicles (BEV) reveals that their performance is strongly dependent upon the primary energy source used to generate the electricity. Generated in a conventional steam power plant fired with coal, the WTW energy consumption is marginally better (8%) than a gasoline-powered vehicle while the WTW greenhouse gas emissions for the BEV is 17% worse than for the gasoline-powered vehicle. On the other hand, electricity generated in a state-of-the-art, high efficiency combined cycle power plant fired with natural gas reduces energy consumption by 35% and greenhouse gas emissions by 55% compared to the gasoline-powered vehicle. At 130 g CO_2 emissions per kilometer driven, this performance equals the ethanol-fueled hybrid electric vehicle while easily besting

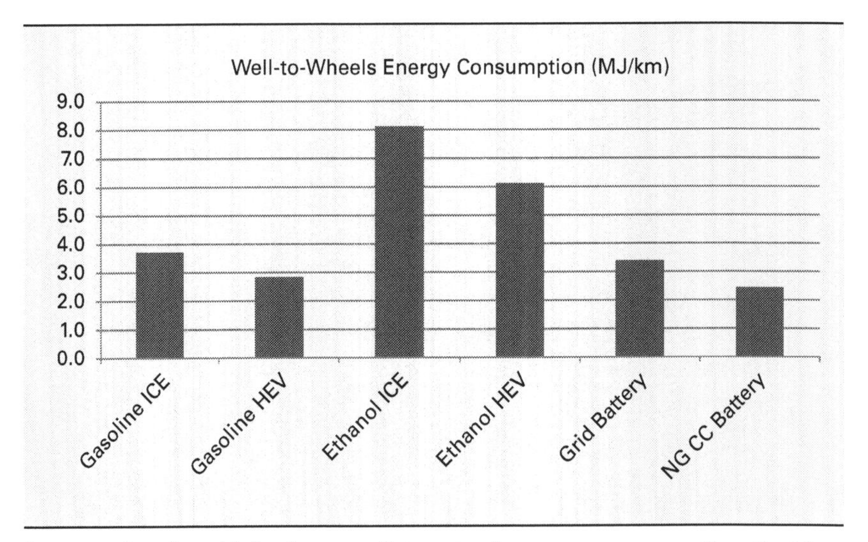

Battery electric vehicles have well-to-wheels energy consumption that is comparable or lower than vehicles powered by gasoline or other alternative fuel options. ICE-internal combustion engine; HEV – hybrid electric vehicle; CNG – compressed natural gas; Grid – electricity from the existing electric grid; NG CC – electricity from natural-gas fired combined cycle power plants. *Source: Data from Reference 5.*

ethanol in terms of WTW energy consumption.

Electricity is also the least expensive energy carrier for transportation in terms of operating cost. Whereas the fuel cost for a typical gasoline-powered internal combustion vehicle today is approximately $0.13 per mile traveled (assuming $3.00 per gallon gasoline), the cost of electricity is only about $0.03 per mile traveled.

Off-setting these favorable attributes is the absence of suitable battery technology for use in long-range vehicles (250 miles between recharging). Although lithium-ion batteries have received considerable publicity for their sevenfold improvement in energy storage capacity per unit weight

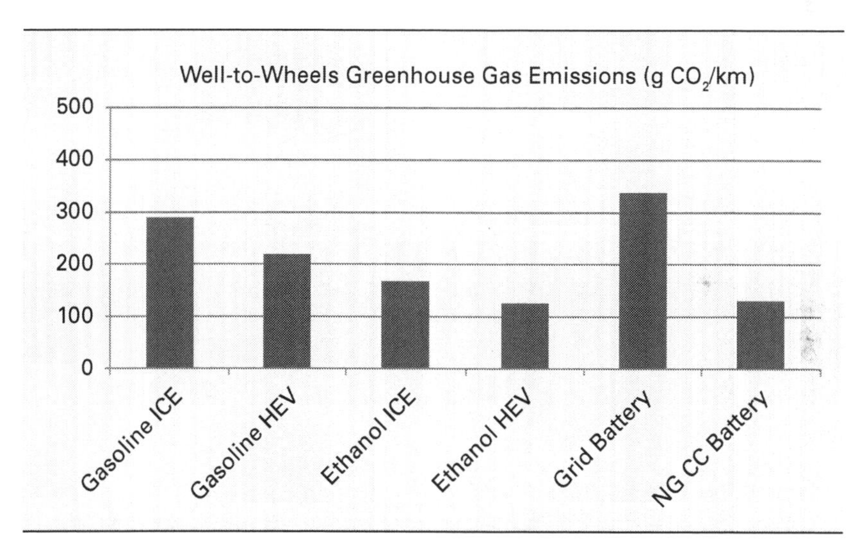

Battery electric vehicles have well-to-wheels greenhouse gas emissions that compare favorably with other low-carbon vehicle options. ICE-internal combustion engine; HEV – hybrid electric vehicle; CNG – compressed natural gas; Grid – electricity from the existing electric grid; NG CC – electricity from natural-gas fired combined cycle power plants. *Source: Data from Reference 5.*

compared to the traditional lead-acid battery, they require a hundred-fold improvement to equal the energy density of gasoline.[20] Even accounting for the much higher efficiency of converting electricity to mechanical power (90%) compared to a spark-ignition internal combustion engine (30%), a BEV would require a battery weighing 15 – 70 times more than a tank of gasoline sized to provide the same driving range (the claims for Li-ion battery technology vary widely). Further exacerbating the short-coming of batteries is the high cost to manufacture them. Whereas gasoline or ethanol can be stored in a simple steel tank costing a few cents per megajoule of energy storage capacity, lithium-ion batteries currently cost $190 to $290 per megajoule of energy storage capacity. For a vehicle with a driving range of 250 miles (417 kilometers), the cost of a gasoline fuel tank would be less than $100 while a Li-ion battery would cost $38,000 to $140,000 to achieve the same range.[21] To get a better appreciation of the actual operating costs, capital costs of the vehicle and total energy costs over the life of the vehicle should be added together (this still neglects the maintenance costs and depreciation cost of the vehicle, which can be appreciable). On the basis of overall vehicle operating cost, a lithium-ion BEV is twice as expensive as a vehicle powered by an internal combustion engine and fueled with gasoline or E85 even when making the most optimistic assumptions about BEVs. Clearly, we are a long way from driving cross-country on battery power. The weight of batteries is likely to prevent their use in military or commercial aviation.

Plug-in hybrids

Hybrid-electric vehicles (HEVs) have become increasingly popular in the U.S. over the last few years, with versions,

Electric vehicles will be an attractive transportation option if a light-weight, low- cost battery can be developed.*

Energy storage technology	Energy cost ($ per mile)	Storage system weight (kg)	Storage system capital cost	Vehicle operating cost ($ per mile)
Gasoline	0.13	30	$68	0.25
E85	0.14	50	$77	0.27
Compressed natural gas	0.08	120	$300	0.29
Lead-acid battery	0.03	3200	$22,000	0.27
Lithium-ion battery	0.03	480-2200	$38,000-$140,000	0.51

Source: DOE EIA and BP

*Cost of gasoline assumed to be $3.00 per gallon. Cost of electricity is based on Reference 22, which excludes transportation fuel taxes likely to be assessed. Storage system sized for 250 mile driving range and assumed to have 200,000 miles useful life. Weight based on full fuel tanks, which is assumed not to affect vehicle range. Vehicle operating cost is sum of energy cost and "per mile" capital cost of storage system (excludes maintenance costs and capital depreciation charges). Li-ion battery data comes from Reference 23 (low range) and Reference 24 (high range).

such as the Toyota Prius, achieving significant sales growth. The standard HEV combines a conventional internal combustion engine (ICE) with a rechargeable energy storage system (RESS) - a rechargeable battery.[25] These vehicles automatically shut down their ICEs and run on the battery when power demand is low, the ICE automatically restarting when power demand exceeds the battery's capability. HEVs consume roughly 30% less fuel than traditional ICE-powered cars, attaining 40-50 miles per gallon (mpg) in both city and highway driving. When battery power runs low it is recharged by a combination of the ICE and kinetic energy produced by the vehicle's brakes. This ability to recharge the battery "on the fly" frees the HEV from the range constraints characteristic of pure electric vehicles.

HEVs are already becoming obsolete due to technological developments in two separate fields. The first of these is the development of the plug-in HEV (PHEV). As the name implies, PHEV batteries are recharged by plugging them directly into the electrical grid via power outlets, allowing the battery to be recharged overnight. The first mass-produced PHEV became publicly available in China in early 2008[26] and in the U.S. in late 2010.[27] Several additional models are expected to be introduced in 2011 and 2012.

The second development is an increase in battery endurance allowing PHEVs to run on pure battery power over significant distances. The Chevrolet Volt, for instance, is capable of running solely on their batteries for up to 35 miles without a recharge (prompting the designation PHEV-35, with the number indicating the vehicle range in miles when operating on battery power). Given that most daily driving is done in a localized area rather than over great distances, this greatly cuts down on fuel consumption as people will only use liquid fuels during intercity travel. Seventy-five percent of Americans, therefore, would be capable of making their regular daily commute without burning any gasoline using a PHEV-35.[28]

To recognize their true potential, however, PHEVs can be combined with biofuels to produce flex-fuel PHEVs. Flex-fuel vehicles (FFVs) are those that can run on blends of ethanol and gasoline containing up to 85% ethanol. When run on the highest blends of ethanol, FFVs can produce up to 70% fewer CO_2 emissions, depending on the source.[29] Creating a flex-fuel PHEV would combine the PHEV's reduced fuel consumption with the FFV's reduced greenhouse gas emissions. An expansion of solar and wind energy sources for the electricity used to recharge the PHEV's battery would further decrease the vehicle's overall CO_2 emissions.

An imposing challenge is the technological constraints on battery storage. The increases in size necessary for increased battery storage capacity carry significant weight increases with them. At a certain point the increased weight (which must be carried by the battery-powered engine) negates the benefits of the increased storage capacity and the larger battery actually decreases the vehicle's battery-only operating range.[30] Virtually all of the major automakers believe they have reached the limits of battery capacity development and now instead focus on decreasing the amount of energy used by the electric engine. Smaller companies devoted to battery production have continued research and early indications are that the type of battery necessary for PHEV60 production is not a technological impossibility. It is currently estimated that PHEV60s produced in the near future will cost $10,000 more to manufacture than HEVs, largely because of the battery expense.[19] If PHEV-60s are to become a popular replacement to traditional gasoline-fueled cars in the future, this price will need to drop.

Despite these challenges, hybrid vehicles remain an attractive approach to reducing both fuel consumption and greenhouse gas emissions. Mass adoption of a PHEV60 that relied on wind or solar power to charge its battery and biofuel to power its ICE would significantly reduce U.S. dependency on foreign crude oil and the amount of greenhouse gases emitted by the transportation sector.

A study by the National Research Council[31] estimates a PHEV-10 battery back will cost about $3,300 while a PHEV-40 battery pack will be $14,000. The difference in retail value between the first PHEV-40 to be introduced to the U.S. market, the Chevrolet Volt, and the average new non-PHEV car is approximately $20,000, not accounting for federal tax credits.[32] Li-ion battery technology has greatly improved the prospects for high energy density batteries, but the report does not expect steep declines in the cost of Li-ion batteries as the commercial opportunity has already been exploited for portable electronic devices. It is likely to take several decades before upfront investment costs for batteries is low enough to offset lifetime fuel savings. The study predicts that tens to hundreds of billions of dollars in subsidies will be required if plug-ins are to achieve rapid penetration of the U.S. automotive market, which optimistically might reach 40 million vehicles by 2030 although more realistically the number is likely to be only 13 million. Even so, plug-in hybrid electric vehicles are not expected to significantly impact carbon emissions before 2030 because grid electricity used to recharge the vehicles is heavily reliant on power plants that burn fossil fuels, particularly coal.

6

Why are we producing grain ethanol and biodiesel?

Brief history of ethanol

Ethanol is the natural product of the fermentation of sugars by yeast and bacteria, a process that was harnessed by human beings millennia ago for the production of intoxicating beverages. Ethanol's use as fuel in internal combustion engines is more recent. Henry Ford's first automobile, the "Quadricycle," was designed to run on ethanol. Ford's Model T – the first mass produced car – was also the world's first flex-fuel vehicle, using an engine that could run on ethanol, gasoline, or a blend of the two.[1] Ethanol's popularity as engine fuel was unable to overcome the combined effects of competition provided by an influx of cheap oil and the era of Prohibition, during which time ethanol producers were frequently accused of collaborating with bootleggers who were illegally selling ethanol beverages in the U.S. Ethanol subsequently disappeared from the American economy for several decades as a result.

During the interim other countries continued to experiment with and develop ethanol, most notably Brazil. Whereas the U.S. had primarily used corn to produce ethanol, Brazil focused on sugar cane, one of its major crops. The first Brazilian government-mandated ethanol blend was instituted in 1931. Germany's unrestricted submarine warfare campaign reduced petroleum imports to Brazil during World War II, giving a boost to its indigenous ethanol market with ethanol production increasing from 26,000 gallons in 1933 to 20 million gallons in 1945.[2] Following the end of the war inexpensive petroleum flooded the country and Brazil saw a decline in ethanol use similar to that witnessed in the U.S. during Prohibition.

The shift away from ethanol was reversed in both Brazil and the United States following the oil embargo by the Organization of Petroleum Exporting Countries (OPEC) against the U.S. and its allies in 1973. The resulting gasoline shortages caused many governments to explore methods for reducing oil consumption, sparking a renewed interest in alternative fuels such as ethanol.

In 1975 Brazil launched an intensive ethanol development program named "PROALCOOL." After overcoming initial resistance from a domestic automobile industry opposed to any major design modifications, the country embarked on large-scale ethanol production. By 1984 94% of passenger cars were fueled by pure ethanol.[3] During the late 1980s, however, Brazil experienced one of the dangers of relying on a single biofuel feedstock as sugar prices underwent record gains, causing the number of vehicles running on neat ethanol to fall to 60% by 1989 and 1% by 1996.[3] This rapid decrease was matched by a correspondingly impressive increase in the number of Brazilian flex-fuel vehicles. Ethanol development in Brazil has progressed to a point where enough is domestically produced annually to provide 40% of the country's transportation fuel supply[4] while still allowing 350 million gallons to be exported annually.[5]

The U.S. government's response to the petroleum embargo did not share Brazil's intensity for utilizing biofuels as an alternative to gasoline. Instead, biofuels were only one part of a multi-pronged effort to reduce reliance on imported petroleum. This effort relied heavily on improving the fuel efficiency standards of automobiles and demonstrating technology for converting coal into transportation fuels. Ethanol was encouraged by the Federal Energy Tax Act of 1978 which granted a $0.40 per gallon subsidy for blending 10% volume of ethanol with gasoline. Legislation in subsequent decades increased this subsidy to as much as $0.60 per gallonbefore completely rescinding it in late 2011.[6] Gasohol, as it was known, encountered problems related to poor compatibility with the existing fuel infrastructure of the United States. The most troubling of these problems were eventually overcome but not before the name "gasohol" became a marketing liability. The fuel blend was eventually redubbed E10. Nevertheless, interest in fuel ethanol waned during the 1980s when the real price of petroleum dropped to historically low levels. Iraq's invasion of Kuwait in 1990 and U.S. military intervention in the conflict renewed interest in decreasing U.S. dependence upon imported petroleum from the Middle East. The 1992 Energy Policy Act mandated the federal government's vehicle fleet to use flex-fuel vehicles capable of operating on E85.

Brazil's 20-year head-start on the U.S. in commercializing biofuels and sugar cane's superior energy efficiency relative to corn as an ethanol feedstock caused some policymakers to fear that America's younger ethanol industry would have difficulty competing with Brazilian ethanol. Brazil's sugar cane ethanol is significantly cheaper to produce than America's corn ethanol. In an attempt to protect its indigenous grain ethanol industry the U.S. has imposed a $0.54 per gallon tariff on imported ethanol in 1980.[7]

The terror attacks of September 11, 2001 and the

subsequent revelation that most of the hijackers were citizens of Saudi Arabia, one of America's largest oil suppliers, caused significant apprehension over America's dependence on the Middle East for oil. Increasing demand by China and India and chronic widespread violence in oil producing countries such as Iraq, Nigeria, and Libya caused the real price of oil (and subsequently gasoline) to quadruple between 1988 and 2008.

The 2006 discovery that the fuel oxygenate MTBE, used as a gasoline additive to reduce air pollution emissions from automobiles, was contaminating groundwater led to its banning by several states and subsequent replacement by ethanol. The increasingly hostile behavior by the leaders of Venezuela and Russia towards the U.S. and its allies has raised concerns that these two petroleum-rich nations may resort to "energy blackmail" by withholding energy supplies from U.S. and European markets in the event of disagreements on international policy, much as Russia withheld natural gas from Europe in late 2008. Brazil's achievement of energy self-sufficiency in 2006 relies on a combination of sugar cane-based ethanol and deep-sea oil drilling, which proves that it is possible even for countries without Saudi Arabian-sized oil reserves to become self-sufficient.

In 2006, when ethanol prices peaked, almost 450 million gallons of Brazilian ethanol was imported into the United States.[8] After peaking at $0.60 per gallon in 1984, federal subsidies were gradually reduced to $0.54 in 1990, $0.51 in 1998, and $0.45 in 2008. By the time the subsidy was eliminated in 2011, ethanol had become sufficiently profitable to make federal support superfluous. A thirty year tariff on importation of ethanol also came to an end in 2011, ending a major barrier to improved trade and diplomatic relations between the United States and Brazil. During his first visit to

Washington after the election of President Obama, Brazilian President Luiz Inacio Lula da Silva had denounced the U.S. tariff on Brazilian sugar cane ethanol as counterproductive in efforts to reduce greenhouse gas emissions. Brazil also threatened litigation at the World Trade Organization to remove the ethanol tariffs. Although eager to improve relationships with Brazil, President Obama also strongly supported domestic production of ethanol. During his 2008 presidential campaign President Obama expressed support for corn-based ethanol production while stating that cellulosic ethanol is the "future of biofuels."[9]

Steps taken by the U.S. Federal Reserve to help the U.S. economy recover from the Great Recession via quantitative easing (QE) inadvertently helped make the tariff obsolete, easing its expiration. One effect of QE has been the devaluation of the U.S. dollar, which in turn has caused the Brazilian real to appreciate against the dollar. The combination of QE, a boom in the price of commodities (of which Brazil is a major exporter), and high Brazilian interest rates caused the value of the Brazilian real to increase by 60% against the U.S. dollar between 2009 and 2011.[10] In a reverse of the historical norm Brazilian sugar cane ethanol has become more expensive on an exchange rate basis than U.S. corn ethanol, so much so that Brazil has become a significant importer of U.S. ethanol.[11] This shift caused policymakers to question whether U.S. corn ethanol still required protection from its Brazilian counterpart. In June 2011 the U.S. Senate voted to end the corn ethanol tariff and subsidy.[12] Although corn and ethanol producers initially opposed this change, eventually they relented and both the tariff and subsidy were allowed to expire on December 31, 2011. As long as demand for fuels remains strong, ethanol is expected to survive this policy shift.

Turning corn into ethanol

Corn ethanol plants are of two types, categorized according to how the grain is prepared for fermentation: dry grind and wet milling.[13] Dry grind plants use the whole kernel as fermentation feedstock while wet milling plants soak the grain with water and acid before mechanically grinding it to separate the corn germ, fiber, gluten, and starch components. The capital investment for dry grind is less than that for a comparably sized wet-milling plant. However, the higher value of its co-products, greater product flexibility, and simpler ethanol production can make a wet-milling plant a more profitable investment.

The dry-grind process consists of four major steps: pretreatment, cooking, fermentation, and distillation.[14] Pretreatment consists of grinding the corn kernel into "meal", which is mixed with water, enzymes and ammonia. This mixture ("mash") is then "cooked" to reduce bacteria levels. After cooling, the mash is sent to the fermenter where it remains for 40 hours or more. The beer resulting from fermentation consists of a mixture of ethanol and stillage. Energy intensive distillation of the beer is necessary to separate water and solids from the ethanol and achieve maximum concentrations of 95 percent. This is followed by further purification to 99.5 percent ethanol using molecular sieves.

The fibrous residue remaining upon completion of fermentation is recovered from the base of the beer stripping column, mixed with yeast and other unfermented residues, and dried to a co-product known as distillers' dried grains and solubles (DDGS). This co-product, containing about 25 wt-% protein and residual oil, is a valuable feed for cattle. Profitability of a corn-to-ethanol plant is strongly tied to the successful marketing of DDGS.

Most modern dry milling plants produce more than 2.7 gallons of ethanol per bushel of corn processed with some recent plants exceeding 2.9 gallons.[15] Yields of co-products per bushel of corn are 17-18 lb (7.7-8.2 kg) of DDGS and 16-17 lb (7.3-7.7 kg) of carbon dioxide evolved from fermentation, the latter of which can be sold to the carbonated beverage industry.[14] As a rule of thumb, the three products are produced in approximately equal weight per bushel.

The cost to build a 40 million gallon per year (GPY) dry grind corn ethanol plant was about $47 million in 2006.[16] The intensive building boom in the biofuels industry over the following two years drove up labor, cement, and steel prices to such an extent that the cost of a new plant doubled in that short period. By 2008 the escalating price of corn and the on-set of a global recession slowed the grain ethanol boom, which eased construction costs. Larger plants capture the so-called "economies of scale," which proportionally reduce plant construction costs although it becomes more difficult to secure sufficient stocks of local grain to keep them operating. For this reason, new plants today tend to be 100 million gallons in capacity or slightly larger.

The cost of producing grain ethanol is strongly dependent upon the price of corn and the efficiencies of various operations within the manufacturing plant, which vary from company to company. When corn prices were around $2 per bushel, the cost of producing grain ethanol through dry grind operations was about $0.90 per gallon.[18] The purchase of corn grain accounts for most of this cost. A 50% increase in the cost of corn raises ethanol production costs to over $1.15 per gallon. Since ethanol has only two-thirds the energy content of gasoline, this is equivalent to gasoline costing $1.74 per gallon.[19] Such a cost increase is not

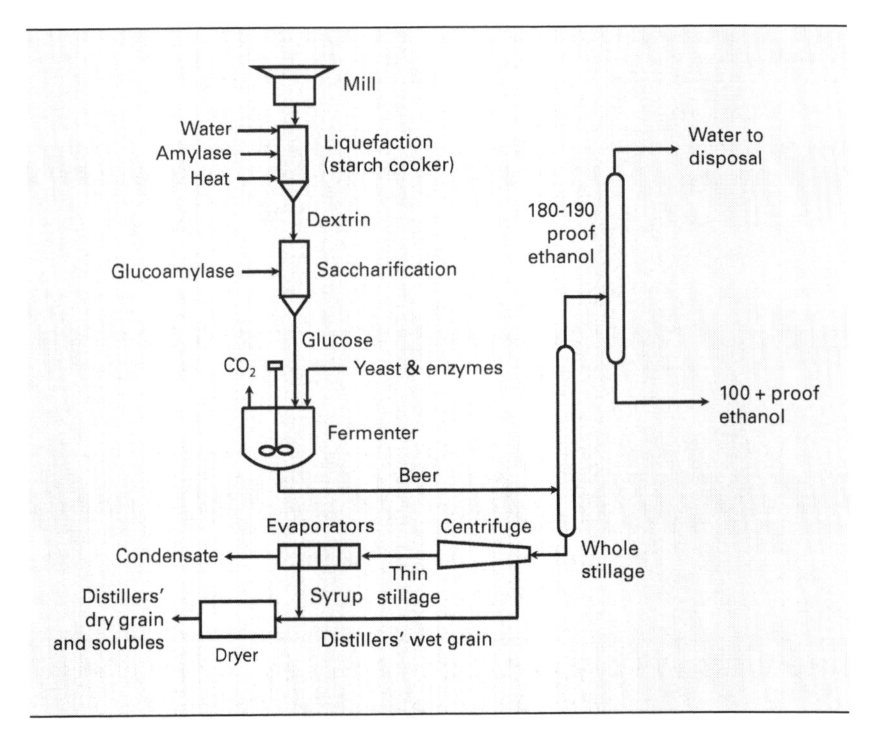

Steps in dry grinding corn to produce ethanol. *Source: Reference* [17].

beyond the realm of possibility: increasing volatility of corn prices has taken it as high as $7 in 2008.

Wet milling has the advantage that it separates plant components into carbohydrate (starch), lipids (corn oil), a protein-rich material (gluten), and fiber (hulls). This gives an ethanol facility access to higher value markets as well as provides flexibility in the use of starch as a food product or in the production of fuel ethanol.

Wet milling consists of a series of steeping, grinding, and separation processes that recovers enriched fractions of oil, protein, starch, and fiber. [20] It begins by soaking the corn kernels in a dilute solution of sulfuric acid for 24-36 hours.

This steeping process swells and softens the corn kernels. Some of the protein and other compounds are dissolved in the resulting corn steep liquor, which represents an inexpensive source of nitrogen and vitamins. After separating the corn from the steep liquor, the wet kernels are coarsely ground to release the hull and germ from the endosperm. Hydrocyclones or screens separate the germ from the rest of the components. After drying, oil is extracted from the germ using either a solvent or screw press, leaving an oil cake. The hull and endosperm pass through rotating disc mills that grind the endosperm into fine fractions of starch and gluten while the hull yields coarser fiber particles, which can be screened out from the finer fractions. Centrifugal separators separate the lighter gluten from the starch.

The starch can be used directly as a food product or for industrial manufacturing processes, especially papermaking. The starch can also be converted to monosaccharides for production of food or fuel, depending on relative market demand. Saccharification by amylase enzymes yields corn syrup, a glucose solution that can be directly fermented to fuel ethanol. Alternatively, when treated with isomerase enzymes the glucose is partially converted to fructose to yield a liquid sweetener known as high fructose corn syrup (HFCS). In plants that can alternate between fuel ethanol and HFCS production, relatively more ethanol is produced in the winter while relatively more HFCS is produced in the summer. The gluten product, known as corn gluten meal, contains 60% protein and is used primarily as poultry feed. The fiber from the hulls is combined with other byproducts, such as the oil cake, steep water solubles, and excess yeast from stillage, dried and sold as corn gluten feed. Containing 21% or more of protein, it is primarily used as feed for dairy and beef cattle.

A typical wet milling plant will produce 9.5-9.8 L (2.5-2.6 gal) of ethanol per bushel of corn processed.[21] Yields of other co-products per bushel of corn are 0.7 kg (1.7 lb) of corn oil, 1.4 kg (3 lb) of corn gluten meal (60% protein), 5.9 kg (13 lb) of corn gluten feed (21% protein), and 7.7 kg (17 lb) of carbon dioxide. Like dry milling, the three wet milling products of ethanol, feed, and carbon dioxide are produced in approximately equal weight per bushel, with each accounting for approximately one-third of the initial weight of the corn. Production costs for ethanol from a wet milling plant are expected to be moderately lower than for a dry grind plant because of the additional value of its co-products.

Turning sugar cane into ethanol

The sugar cane ethanol process consists of four major steps: milling, filtration/evaporation, fermentation, and distillation.[21] Milling consists of chopping and shredding the sugar cane with revolving blades, with the intent of separating the sugar cane juice from the bagasse, the fibrous part of the plant. Bagasse can be used as boiler fuel, containing 1/3 of the total energy found in sugar cane, and is used by Brazilian ethanol plants as their primary source of electricity as well as heat for evaporation and distillation processes in the plant. After the bagasse has been separated the juice is filtered multiple times, pasteurized, and evaporated until a mixture of sugar and molasses remains. After the sugar is removed the molasses is combined with yeast and allowed to ferment for up to twelve hours, producing a liquid similar to wine with an alcohol content of up to 10 percent. Finally, this liquid is distilled in the same manner as corn ethanol to produce 100% ethanol. While the corn ethanol process

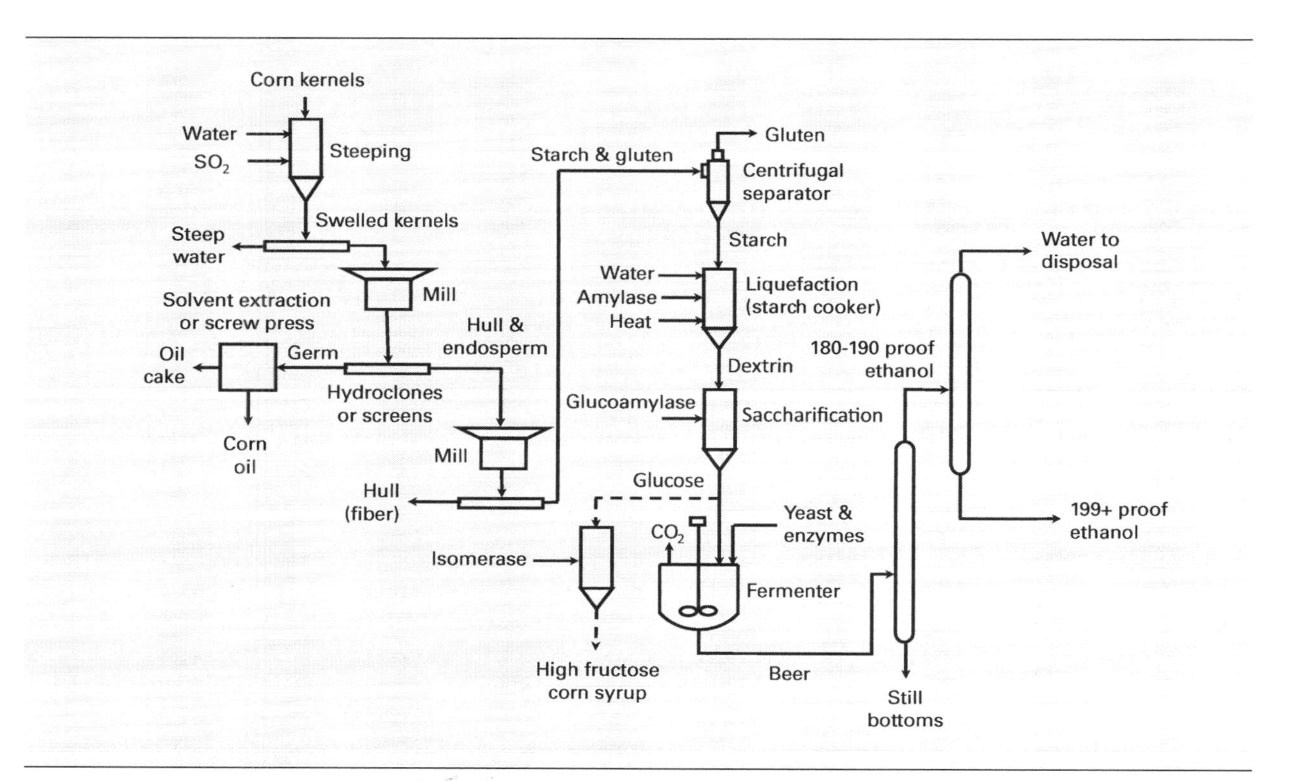

Steps in wet milling corn to produce ethanol. *Source: Reference 18.*

is similar to beer brewing, the sugar cane ethanol process is similar to wine vinting, using sugar cane instead of grapes.

The cost of building a modern sugar cane ethanol plant capable of producing 50 million gallons per year was $150 million in 2008[22] with half of the processed sugar cane being used to produce sugar and the other half to produce ethanol.[22] These plants produce roughly 20 gallons of ethanol per ton of sugar cane processed. The cost of production largely depends on the price of sugar cane, which averaged $10.40 per ton in 2005, or $0.54/gallon. After including investment costs (which vary depending on the original plant costs) and transportation costs under "free-on-board" contracts (under which the producer is liable for all costs until the ethanol is delivered to the suppliers), the ultimate

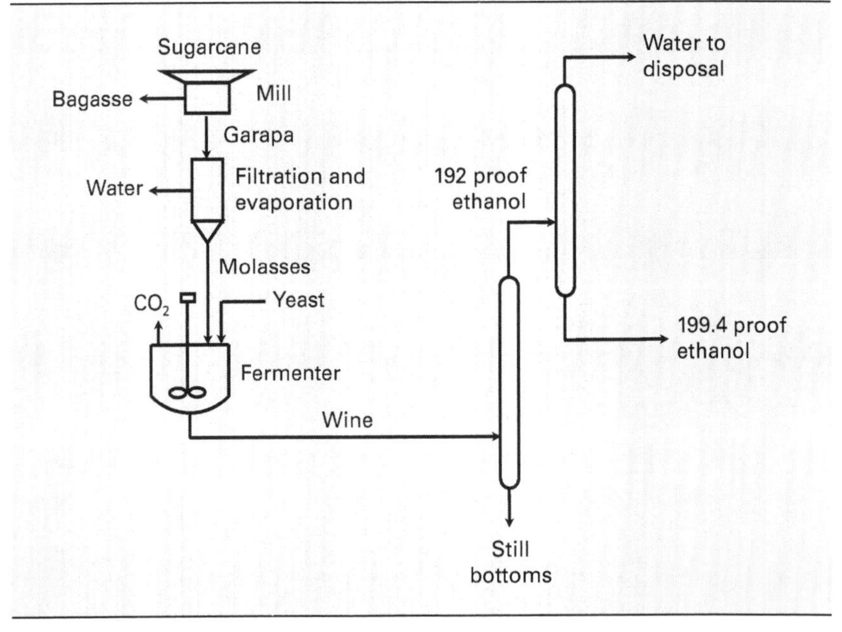

Steps in converting sugar cane into ethanol. *Source: Author.*

cost of production varies from $0.63 - $0.76 per gallon of ethanol, which is 20 -80% less expensive than grain ethanol. Because a gallon of ethanol contains only 67% of the energy that a gallon of gasoline contains, this cost is more accurately portrayed as $0.95-$1.13 per gallon of gasoline equivalent, making it competitive with gasoline at $40-$50 per barrel prices.

The ability to use bagasse, the plant fibers that remain after syrup has been pressed from sugar cane, as boiler fuel makes ethanol from sugar cane relatively fossil-fuel efficient. Whereas the net energy balance - the ratio of the energy in a unit of ethanol divided by the fossil fuel energy required to produce it - for corn-based ethanol is 1.3, the balance for sugar cane-based ethanol ranges between 8.2 and 10, due in large part to the energy self-sufficiency of many Brazilian ethanol plants.[23] This use of bagasse also contributes to a roughly 90% decrease in greenhouse gas emissions resulting from using ethanol instead of gasoline.

Brief history of biodiesel

Biodiesel, manufactured from lipids, including animal fats and plant oils, is a substitute for petroleum-based diesel fuel. Rudolf Diesel, inventor of the engine that bears his name, originally conceived his engine to burn powdered coal although he found better success with petroleum-derived liquids. After the French government successfully burned pure, unprocessed peanut oil in a diesel engine at the 1900 World Fair in Paris, Diesel became interested in lipid-based fuels and declared that fuels based on vegetable oils could become as important as petroleum-based fuels.[23]

Fuels based on vegetable oils faced two serious constraints: high viscosity and competition with an abundance of cheap crude oil. The high viscosity of vegetable oil requires it to be heated before it can be reliably burned in internal combustion engines where it would otherwise foul fuel ports and engine cylinders. While technologically feasible, this additional step required modifications to the engine. The additional cost makes it difficult for the modified engines to compete with engines powered by diesel fuel or gasoline.[24] Crude oil shortages during World War II and OPEC's 1973 oil embargo resulted in brief periods of rejuvenated interest in vegetable oil. Today, methyl ester manufactured from vegetable oil and commonly known as biodiesel has supplanted the direct use of vegetable oil as engine fuel.

Conversion of vegetable oil into biodiesel by a process known as transesterification dates to 1853,[24] fifty years before the invention of the diesel engine. It was not until 1937 that a Belgian scientist first patented transesterification.[25] Crude oil shortages during World War II resulted in limited use of biodiesel in many places around the world, although this promptly disappeared after the cessation of hostilities in 1945.

Biodiesel witnessed a resurgence starting in the late 1970's at which time pure vegetable oil's limitations became apparent even when precautions, such as heating the oil to reduce its viscosity, were taken. Several academic studies indicated that most of these problems could be eliminated through transesterification of the vegetable oil. South Africa was the first country to perfect this process, spurred on by a world-wide embargo against exporting petroleum to South Africa as a response to its Apartheid policies. By 1989 South Africa had built the world's first industrial biodiesel plant, using rapeseed as its feedstock. Austria purchased the South African technology and by the 1990s the Czech Republic, France, Germany, and Sweden were all host to their own biodiesel plants.[26]

U.S. research into biodiesel began in earnest in 1982 and its development became widespread over the course of the next decade when soybean producers, recognizing an opportunity to expand markets for soybeans, became leading advocates of biodiesel. What began in 1992 as a small group of soybean farmers known as the "National Soy Fuels Advisory Committee" quickly transformed into the influential "National Biodiesel Board" (NBB) over the following years. This organization, using the profits of its members, has financed much of the research into biodiesel that has since occurred.[26]

Turning lipids into biodiesel

Lipids are a large group of hydrophobic, fat-soluble compounds produced by plants and animals for high-density energy storage.[26] Triglycerides of fatty acids, commonly known as fats or oils depending upon their melting points, are among the most familiar form of lipids and are widely used for the production of diesel fuel substitutes. Fats, which are solid or semi-solid at room temperature, have a high percentage of saturated acids, whereas oils, which are liquid at room temperature, have a high percentage of unsaturated acids. Plant-derived triglycerides are typically oils containing unsaturated fatty acids, including oleic, linoleic, and linolenic acids.

Suitable feedstocks include soybean, sunflower, cottonseed, corn, groundnut (peanut), safflower, rapeseed, waste cooking oils, and animal fats.[27] Waste oils or tallow (white or yellow grease) can also be converted to biodiesel. A wide variety of plant species produce triglycerides in commercially significant quantities, most of it occurring in seeds.

Average oil yields range from 150 L/ha for cottonseed to 814 L/ha for peanut oil although intensive cultivation can double these numbers.

Extraction of seed oil is relatively straightforward. The seeds are crushed to release the oil from the seed. Mechanical pressing is used to extract oil from seeds with oil content exceeding 20% and solvent extraction is required for seeds of lower oil content. The residual seed material, known as meal, is used in animal feed.[28]

Transesterification describes the process by which triglycerides are reacted with methanol or ethanol to produce methyl esters and ethyl esters, respectively, along with the co-product glycerol. One triglyceride molecule reacts with three methanol molecules to produce one molecule of 1, 2,

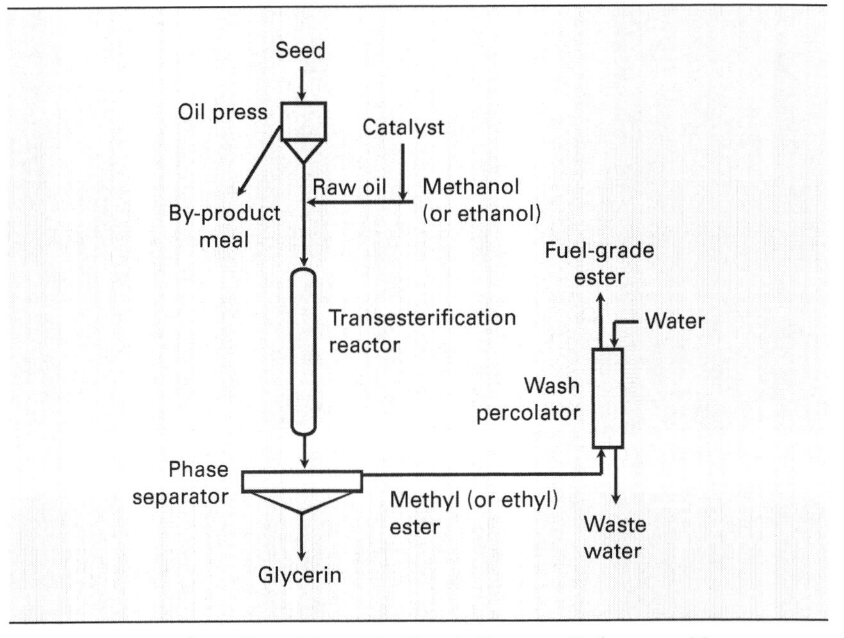

Steps in converting oilseed into biodiesel. *Source: Reference 29.*

3-propanetriol, which is known as glycerol, and three ester molecules.[29]

The fuel properties of biodiesel are very similar to petroleum-based diesel. The specific gravity and viscosity of biodiesel are only slightly higher than for diesel while the cetane numbers and heating values are comparable. Significantly higher flash points for biodiesel represent greater safety in storage and transportation. Biodiesel can be used in unmodified diesel engines with no excess wear or operational problems. Tests in light and heavy trucks showed few differences other than a requirement for more frequent oil changes because of the build-up of unburned ester fuel in engine crankcases.[30]

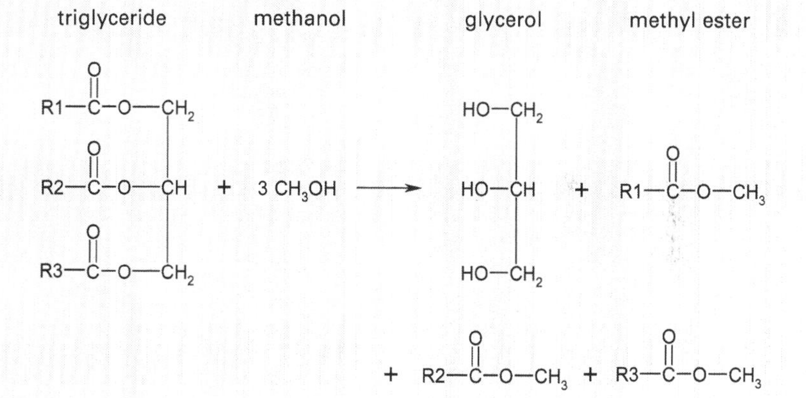

triglyceride | methanol | glycerol | methyl ester

$R1-C(=O)-O-CH_2$ / $R2-C(=O)-O-CH$ / $R3-C(=O)-O-CH_2$ + $3\ CH_3OH$ → $HO-CH_2$ / $HO-CH$ / $HO-CH_2$ + $R1-C(=O)-O-CH_3$ + $R2-C(=O)-O-CH_3$ + $R3-C(=O)-O-CH_3$

Transesterification: Chemistry of converting triglycerides into methyl esters. R1, R2, and R3 represent long carbon chains known as fatty acids.
Source: Reference 30.

7

Why are we developing advanced biofuels?

First-generation biofuels, which includes grain ethanol and biodiesel, have been the subject of intense debate in the last several years, garnering both strong support and vocal criticism. While proponents of so-called first generation biofuels have often oversold "renewable fuels" as solutions to a multitude of the world's energy and environmental problems, critics of biofuels have zealously demonized them as both a "crime against humanity"[1] and an "environmental disaster."[2] Most thoughtful people would agree that the solution to our energy and environmental problems will not be achieved by this kind of political posturing. Instead, we need to acknowledge that the path forward is neither smooth nor certain. It will involve adoption of transitional technologies that, although far from perfect, provide a bridge to a more sustainable energy future.

The transition to renewable energy begins by demonstrating that there are alternatives to fossil fuels. Developers of first generation biofuels have been successful in that respect. It is inevitable that first generation biofuels, like any new technology, will have shortcomings, but the solution is not to abandon their further development. Such essentials of modern civilization as electricity and medicine would not exist today if perfection had been demanded upon their first introduction. Instead, lessons learned from these problems help guide the development of advanced biofuels technologies and the formulation of more effective energy, agricultural, and environmental policies. This chapter examines the controversies surrounding first generation biofuels and sets the stage for exploring advanced biofuels.

Are ethanol and biodiesel poor substitutes for gasoline and diesel fuel?

When people are asked why they do not use ethanol in their cars, a common response is that their mechanic told them it would destroy their engines. Considering the distinct physical and chemical properties of gasoline and ethanol, it is not surprising that the first commercial trials experienced problems including corroded fuel tanks, clogged fuel filters, and leaky gaskets and seals. These fuel incompatibility problems were eventually diagnosed and corrected by limiting the amount of ethanol blended in gasoline to 10% by volume, designated as E10, and changing specifications for susceptible components in automobile fuel systems and engines. However, complaints remain, which hampers more widespread adoption of ethanol.

Oxygen content is responsible for many of these differences. Gasoline consists of pure hydrocarbons and contains no oxygen unless deliberately added as "fuel oxygenate." Ethanol contains 35% oxygen by mass. Thus, some of the complaints about ethanol can never be fully overcome as long as our infrastructure and performance expectations are based on hydrocarbon fuels.

The heating value of a fuel is determined by the amount of heat released upon burning the fuel in oxygen. The oxygen in ethanol makes it "partially oxidized" compared to hydrocarbons that have similar numbers of carbon and hydrogen atoms. As a result, the heating value of a gallon of ethanol is only 66% that of a gallon of gasoline. Accordingly, ethanol is expected to get only 66% of the driving range of a gasoline when burned in an internal combustion engine designed for gasoline. This difference is hardly noticeable in automobiles fueled with E10 but it results in a significant loss in driving range for a vehicle operated on E-85, an 85% blend of ethanol in gasoline. Instead of a 450 mile range, a car operating on E-85 might require refueling after only 300 miles. Although a range of 300 miles for fuel cell vehicles or battery electric vehicles would be hailed as the technological achievement of the century, many commentators find it unacceptably low for biofuels.

Fuel economy depends on many complex interactions between fuel and oxygen molecules in the hot, reactive environment of an engine, which offers opportunities to increase the range of ethanol-fueled vehicles.[3] For example, the higher octane number for ethanol compared to gasoline (109 vs. 91 to 101) allows engines to be designed to run at higher compression ratios, which improves both power and fuel economy. Estimates for efficiency improvements in engines optimized for ethanol instead of gasoline are 15% to 20%, resulting in a driving range approaching 80% of that of a gasoline-powered vehicle.[4] Instead of a 450 mile driving range, such an ethanol vehicle would have to refuel every 360 miles.

A more serious problem is the interaction of water with ethanol. Hydrocarbon fuels like gasoline or aviation fuel are hydrophobic; that is, they repel water molecules. Ethanol is hydrophilic and readily absorbs water. Although water can be a problem in any fuel delivery system, the affinity of ethanol for water means it is more likely to develop into a serious problem. Water contaminating a storage tank or pipeline is readily absorbed by ethanol, resulting in separation of the fuel blend into a lower water-rich layer and an upper hydrocarbon-rich layer, which interferes with proper engine operation. Water contamination is a problem that has not been fully addressed by the refining, blending, and distribution industries; thus transportation of ethanol-gasoline blends in pipelines is not permitted in the United States and long-term storage is to be avoided.[3]

Ethanol is a powerful solvent, which is responsible for many of the early problems arising from its substitution for gasoline. Over time gasoline can deposit "varnish" or "sludge" in fuel tanks, which normally does not present problems. When ethanol was introduced to the fuel tanks of older automobile, it softened and released accumulated deposits, which plugged fuel filters. Many of the rubber gaskets and seals in fuel systems and engines were designed for the relatively unreactive compounds found in hydrocarbon-based gasoline. Ethanol aggressively attacked these materials, sometimes necessitating expensive repairs to automobiles.

The hydrophilic nature and high solvent power of ethanol are fundamentally incompatible with our present fuel infrastructure. This has been effectively addressed by capping the amount of ethanol blended with gasoline to 10% volume and permitting as much as 15% ethanol by volume in newer vehicles. However, this cap also establishes a "blend wall" that limits the total amount of ethanol that can be used as transportation fuels in the United States. If every gasoline-powered vehicle in the United States were powered by E10 we would in theory replace 9 billion gallons of the 140 billion gallons of petroleum-based gasoline that is annually consumed with 13 billion gallons of ethanol (the larger amount of ethanol is the result of it having

70% lower volumetric heating value compared to gasoline). In reality the total amount of ethanol displacing gasoline will be closer to 12 billion gallons due to area and seasonal restrictions.[5] The Energy Independence and Security Act of 2007 mandates the production of 15 billion gallons of grain ethanol alone and another 16 billion gallons of "cellulosic biofuels," which may or may not be ethanol. While the recent national increase to E15 alleviates this problem somewhat, a low utilization rate by the public ensures that the blend wall continues to loom as a barrier to expansion of the ethanol industry.

An obvious solution is to go to substantially higher blends of ethanol in gasoline. If every gasoline-powered vehicle in the United States was fuel flexible and able to burn E-85, the amount of grain or cellulosic ethanol that could be absorbed in the U.S. fuels market would only be limited by overall transportation fuel consumption. But such a change requires massive investment in new infrastructure, including the production of more flex fuel vehicles and addition of E85 pumps at gasoline stations. Although not impossible, it comes at considerable expense and would require several years to implement. The U.S. Department of Energy estimates the cost of modifying all U.S. gasoline stations to be from $3.4 billion to $10.1 billion.[6] Modifying the existing pipeline infrastructure and vehicles to be compatible with E85 would only add to this cost. For this reason ethanol advocates are calling for use of 15% blends of ethanol with gasoline in standard automobiles. This raises old questions about engine compatibility and environmental performance that will also take time to sort out.

Early studies of oxygenated fuels suggested that they were more "clean-burning" than gasoline, reducing emissions of carbon monoxide, benzene, particulate matter, and volatile organic compounds. But these early tests were performed in carbureted engines, which have largely been replaced by computer-controlled, fuel-injected engines able to achieve near complete combustion independent of the oxygen content of the fuel. Thus, oxygenated fuels play a smaller part in fighting air pollution than was once the case. In fact, there is evidence that ethanol-gasoline blends exacerbates emissions of so-called volatile organic emissions (VOCs) from automobiles as a result of increased fuel volatility: although tail pipe emissions of VOCs might decrease marginally, fuel tank emissions increase significantly, from 34% of total VOC emissions from conventional gasoline to 42% of total VOC emissions from gasoline reformulated with ethanol. The problem occurs because the vapor pressure of blends of gasoline and ethanol are higher than from either fuel component alone. The problem is particularly acute for E10 but is substantially ameliorated for higher ethanol blends such as E85.

Oxygenated fuels may affect nitric oxide (NOx) emissions, a precursor to acid rain. Although ethanol contains no nitrogen, high oxygen levels promote thermal NOx formation. On the other hand, the lower flame temperature for combustion of ethanol should favor lower thermal NOx formation. Engine testing has produced inconsistent results but some environmentalists argue that ethanol-fueled automobiles exacerbate NOx emissions. The U.S. Environmental Protection Agency projects that the national renewable fuel standard for 2010 will reduce CO and benzene emissions but increase emissions of hydrocarbons (that is, VOC), NOx, ethanol, and acetaldehyde (an ethanol decomposition product).[7]

Biodiesel has relatively fewer shortcomings as a substitute for petroleum-based fuels in part because it has physical and chemical characteristics very similar to diesel fuel.[8] In fact, the methyl esters that constitute biodiesel closely resemble the long-chain hydrocarbons found in diesel fuel except for the presence of a relatively small amount of oxygen at one end of the hydrocarbon chain and an occasional double bond between carbon atoms in the chain. One of the main complaints about biodiesel arises from these double bonds, which make the fuel susceptible to oxidation and degrades biodiesel that has been stored for long periods of time.[9] Another occasional problem is cold-weather performance of bio-diesel, which can form waxy particles that plug fuel filters or even form a gel that cannot be pumped. Although easily avoided by careful quality control in biodiesel manufacturing plants, occasional problems have been prominently reported in the press.[10]

Is it possible to grow enough corn and soybeans to meet our energy needs?

A number of critics of first generation biofuels have pointed out the obvious in both scholarly[11] and popular press articles:[12] "We can't grow enough corn and soybeans to meet our energy needs." The U.S. corn harvest averaged almost 11 billion bushels between 2001 and 2008.[13] If the whole corn crop was dedicated to ethanol production, modern dry mill ethanol plants could produce a little more than 30 billion gallons of ethanol from it. Ethanol has only two-thirds the heating value of gasoline, which suggests that this amount of ethanol is only equivalent to 20 billion gallons of displaced gasoline. If internal combustion engines were optimized for use of high octane fuel, the driving range would approach 80% of the range for gasoline,[4] and the corn crop could produce the equivalent to 24 million gallons of gasoline. Although this is a lot of fuel, it would only displace 14% to 17% of the 140 billion gallons of gasoline currently used in the United States. The seed industry is projecting increases in average yields from 150 bushel to 300 bushel per acre as a result of plant breeding and biotechnology programs.[14] This would still provide only 28% to 34% of gasoline demand in the United States and would leave no corn for the cattle, swine, or poultry industries or markets for starches and sweeteners. A similar analysis of soybean production indicates that only 6% of U.S. demand for diesel fuel could be met with biodiesel.[11] There simply is not enough farmland in the U.S. to grow sufficient corn and soybeans to eliminate the country's dependence on crude oil. Neither grain producers nor the biofuels industry have ever claimed differently.

The ethanol and biodiesel industries emerged as a result of two previously unrelated supply and demand issues in the United States: excess supply of agricultural commodities compared to world demand and excess demand for petroleum compared to world supply. Advances in agriculture have increased yields of corn and soybeans dramatically over the course of the twentieth century with further improvements expected in the present century. As a result, the U.S. is producing almost twice as much corn crop today as it did thirty years ago on approximately the same amount of farmland. The increase in soybean yields has been a more modest but still impressive 63% in the same time period.

This agricultural bounty has not rewarded the American farmer, or for that matter, farmers anywhere in the world. Between 1983 and 1987 the average price of a bushel of Iowa corn was $2.19.[15] Adjusted for inflation, this is equivalent to $3.75 in 2003 dollars. The average price of Iowa corn between 2001 and 2005, just before the ethanol fueled run-up in corn prices, was only $2.08 in 2003 dollars. Three decades of agriculture progress rewarded Iowa farmers with inflation-adjusted corn prices that were lower by 44%. A similar analysis reveals a 42% decrease in inflation-adjusted prices for Iowa soybeans over the last three decades.[15] Although a success for the cheap food movement, it has hurt farmers. Whereas crop prices have not kept up with inflation, the same cannot be said of farm machinery, land rents, labor, seed, fertilizer, and other inputs to agriculture. Although a farm's profitability is very much dependent on individual management practices, fertility of the land under cultivation, and vagaries of weather, in general a farmer will not start making money from corn until the price is above $3.00 per bushel if they own their land and $4.00 per bushel if they rent their land.[16] This failure to "keep up with inflation" explains why subsidies have been a main stay of American agriculture for many years.

High productivity and low profitability have been problems for American agriculture for many years. In the 1970s the U.S. government promoted grain exports to other nations, most notably to the Soviet Union, as a solution to excess production capacity in the U.S. This policy increased exports at an annual rate of 20% between 1970 and 1980 with the value of agriculture exports increasing from $6.7 billion to $40.5 billion.[17] But a number of factors conspired in the following decade to reverse these favorable conditions including a global recession, political discord between the U.S. and the Soviet Union, and increasing crop harvests in other parts of the world. By the early 1980's, the Congressional Budget Office reported that U.S. agriculture was characterized by "depressed farm prices, declining net trade balance, record reductions in farm acreage, and unprecedented federal outlays for agriculture price supports."[17]

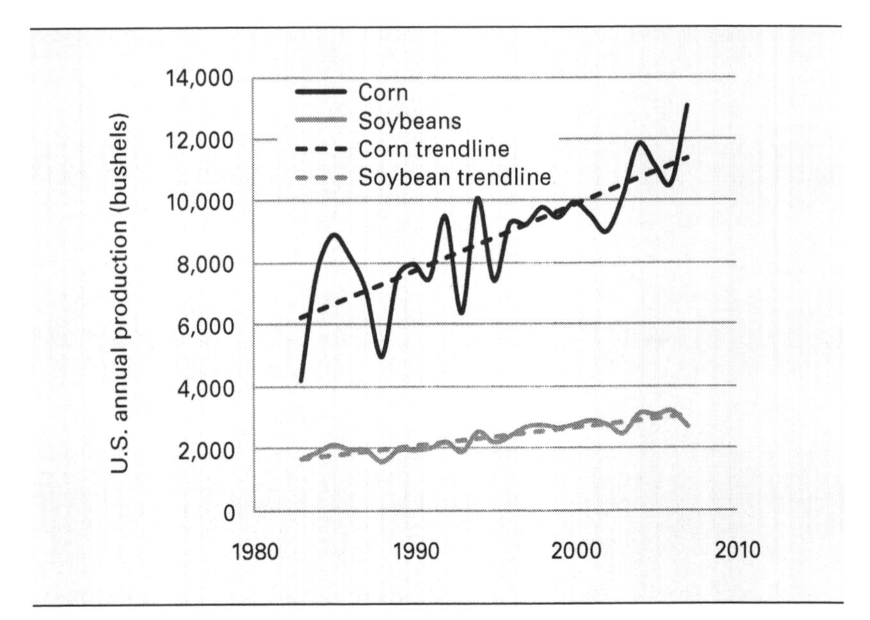

Success of American agriculture: Increasing harvests. Annual production fluctuations around the increasing trend lines reflects both the influence of weather and farmers decisions about whether to expand or contract crop production based on market conditions. *Source: Reference 15.*

The issue has never been whether American farmers can grow enough corn and beans to overcome our national addiction to oil—they cannot. Instead, it has been a matter of whether the excess supply of commodity crops could be used to help displace demand for imported petroleum, to the benefit of both American farmer and consumer. The American experiment with first generation biofuels has shown us the possibilities. The Energy Policy Act of 2005 set in motion an expansion of the ethanol industry, which increased the price of corn, not as an unintended consequence but as the expected outcome of agriculture policy designed to expand markets for agriculture products. Iowa corn prices moved

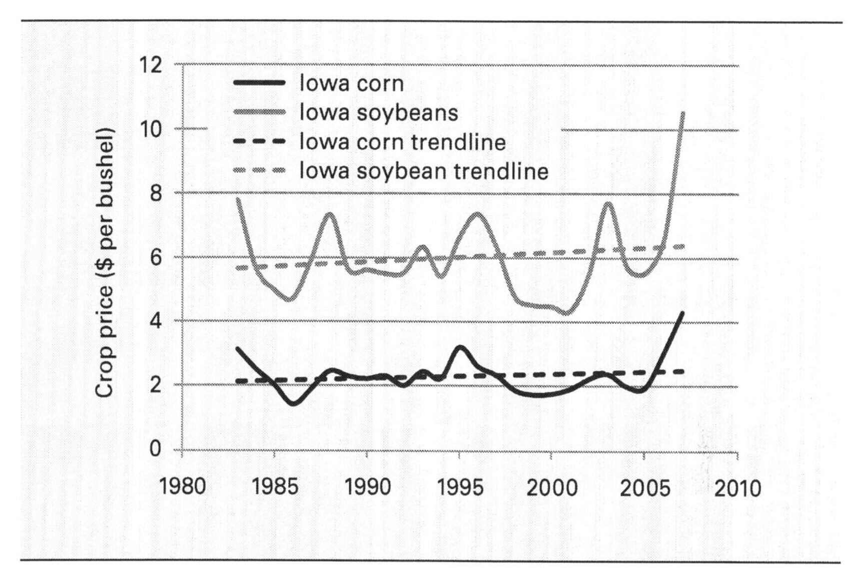

Success of American agriculture: Cheap food for the world. Nominal (not adjusted for inflation) Iowa crop prices barely crept up in the thirty years before the ethanol boom of 2006-2008. Adjusted for inflation, the five-year average (real) price for Iowa corn declined 45% during the last thirty years. *Source: Refereence 15.*

from an economically unsustainable $1.94 per bushel in 2005 to a modestly profitable $4.29 in 2010, which on average provided land-owning farmers net profit of $202 per acre. With Iowa farms averaging 356 acres,[18] an Iowa farmer who owned his land could expect just under $72,000 in net farm income, a relatively modest return for effort in comparison to many other careers in the U.S. On the other hand, a young farmer who was cash renting farmland would only net $9,700 for the same size farm, as a consequence of the current high land rents in Iowa and other states.

Of course, this deliberate expansion of markets for corn and soybeans raises questions about its effect on food prices

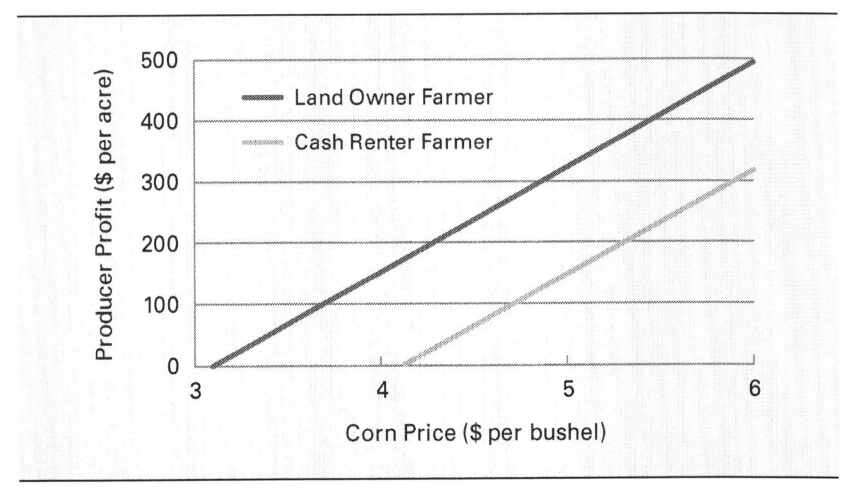

Failure of American agriculture: Lack of profitability. Farmers do not profit from growing corn until prices exceed about $3 per bushel for those that own their farmland and $4 per bushel for those that cash rent their farmland. Historically Iowa corn has been closer to $2 per bushel until 2006. *Source: Reference 16.*

and availability, price supports and subsidies, and environmental impacts. These concerns are explored in the following sections.

Isn't biofuels agriculture responsible for increasing food prices and world hunger?

Between January 2006 and July 2008, the prices of food staples such as rice, wheat, corn and soybeans rose rapidly doubling and even tripling in some cases. Americans saw the price of all kinds of groceries increase. The Grocery Manufacturers Association launched a public relations campaign against ethanol, claiming that rising corn prices were responsible for most of the increase in food prices.[19] The domestic effects of this price inflation in Western countries, while not inconsequential, were dwarfed by the response in developing countries where widespread riots broke out over food distribution.[20] It was not long before various international bodies labeled biofuel production, both corn- and soybean-based, as the culprit. The World Bank declared biofuel production in the U.S. and Europe to be the "main reason" behind the food price increases.[21] It was at the height of this rising panic that a United Nations official declared the use of farmland for biofuel production to be a "crime against humanity."[22] On the face of it, blaming ethanol for food shortages seemed reasonable. Over the last decade the fraction of the U.S. corn crop going into ethanol production rose from 5% to 23%, which suggested that food had been diverted away from food markets. No one could deny that corn prices began to rise about the time the ethanol industry began its rapid expansion.

But the "food vs. fuel" argument is premised on the misperception that corn grown in the U.S. is a major food crop. It is not. Most of the U.S. crop is yellow dent corn (*Zea Mays Indentata*), known by farmers as "field corn."[23] Yellow dent corn is grown mostly for livestock feed, artificial sweeteners (high fructose corn syrup), and ethanol. Only 3% of the U.S. corn crop is grown for human food consumption.[24] These are special food-grade varieties of both yellow and white dent corn, intended for production of corn flakes, corn grits, corn meal, and corn flour. Some countries, such as Mexico, grow a larger portion of their corn crop as food-grade corn to meet demand for corn flour products such as tortillas. In the U.S., though, the production capacity of U.S. agriculture far exceeds domestic and export demands for food-grade corn. Soybeans are also mistakenly considered to be grown primarily as food. Although its lipid content is valuable as cooking oil and as a food ingredient, it represents only 20% of the total weight of soybeans.[25] The rest is soy meal, a co-product of crushing soybeans to extract its oil content, which is mostly used as animal feed.

Most corn and soybean production in the U.S. goes toward livestock feed, sold both domestically and abroad. The United States alone is home to 103 million cattle, 67 million swine, and 9.3 billion chickens and turkeys.[26] This U.S. livestock population of 9.5 billion animals dwarfs the U.S. human population of 311 million and even exceeds the world human population, which stands at 7 billion. Of course, this feed is converted by livestock into protein and fat, which becomes part of our food supply, but the process is very energy inefficient. One hundred calories of energy in the form of corn, if used as livestock feed, ends up as only 12.8 calories in chicken, or 10.4 calories in pork, or 5.3 calories in beef. In comparison, ethanol retains 55 calories of every 100 calories worth of corn used in its production.[27] Just as grossly inefficient energy utilization was economically acceptable in the days of inexpensive fossil fuels, we still accept this large inefficiency in food production because excess agriculture capacity exists in many parts of the world today. If world population growth continues, we will have to consider more efficient ways to provide fuel and food to the world.

Nevertheless, corn used for feed ultimately yields food, however inefficient. If this corn is diverted to processes that yield fuel instead, then the food supply may be negatively impacted.[13] During the last decade there is no evidence of such a shift. In fact, the amount of corn going to feed production has increased by 12% over the last decade and even at the peak of the ethanol boom there was virtually no change in the amount of corn going to livestock

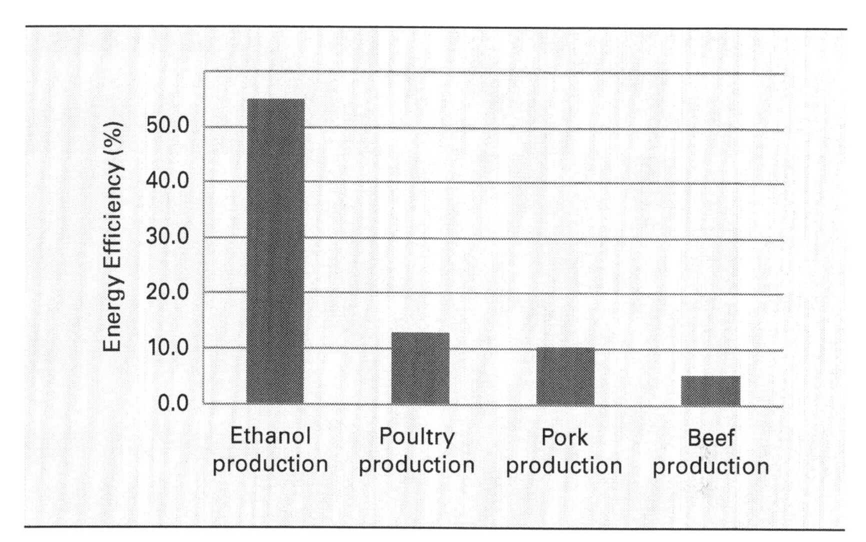

Feed vs. food? Use of corn for livestock production denies the world more food than does ethanol production. *Source: Reference 27.*

production. During the same period exports, which are primarily sold into feed markets, increased 66%. The absence of a shortfall in corn is the result of farmers planting more acres with higher yielding corn varieties – American agriculture is producing 50% more corn than it did a decade ago, meeting the needs for food, feed, and fuel.

And yet, the price of corn went up because of speculation about future demand for corn outpacing supply. At the very least, ethanol critics contend, corn ethanol is largely responsible for the price of food products that rely heavily on corn such as meat, dairy, eggs, soft drinks, and tortillas. In fact, farm products in general contribute only 19% towards the price paid for food by consumers at the grocery store, with transportation costs and the middleman taking the rest.[28] In May 2008, President Bush's Council of Economic Advisors claimed that biofuels production accounted for only 3%

of the 43% global increase in food prices.[29] A 2010 report by the World Bank found that biofuels played a relatively small role in the 2008 increase in food prices, with most of the blame being attributed to high energy prices.[30] Weather, political instability, the lack of transportation and storage infrastructure, high oil prices, excess monetary liquidity, and government policies that discourage agricultural development are among the major factors that determine the price of food around the world.

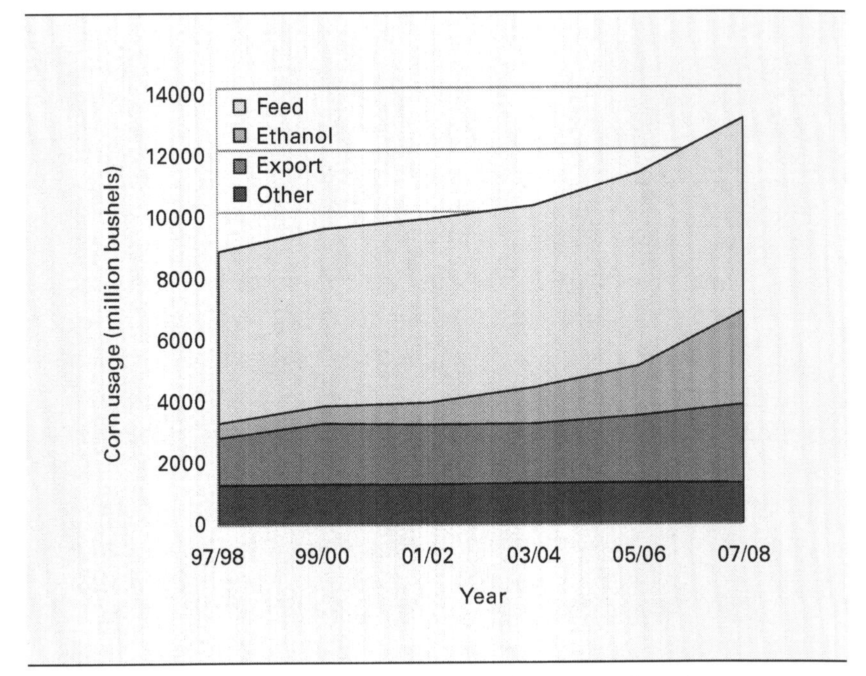

Corn exports actually increased during the time that ethanol was supposedly diverting grain away from world food production and encouraging the conversion of natural ecosystems to agriculture. *Other denotes corn used in food ingredients, industrial products, and consumer products. *Source: Reference 13.*

It did not take long to determine whether the ethanol industry was responsible for price run-ups. After peaking in early July 2008, prices for corn and soybeans, the two most common feedstocks for U.S. biofuel production, began dropping and by December 2008 were essentially back to where they had been two years earlier. The same could be said of most other commodity crops as well as the food price index. And yet ethanol production was at an all-time high, producing almost 10 billion gallons of annual production, which was 100% more than two years earlier. The absurdity of the claim that ethanol production was responsible for food price increases was laid bare and even Scott Faber of the Grocers Manufacturers Association, who led the food

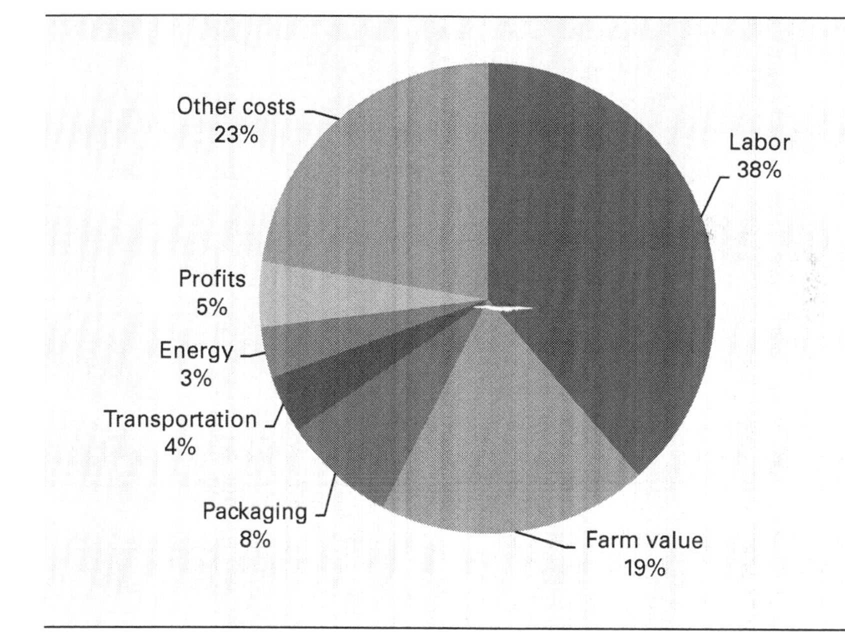

The Price of Food: Farm value contributes on 19% of the total cost of retail food prices. *Source: Reference 28.*

vs. fuel charge against the ethanol industry, was forced to make the following concession:[31]

"Ethanol production is just one of seven sources of commodity price inflation. The rise in global demand, energy prices, speculation, the weak dollar export restrictions and poor weather also contributed to the surge in corn prices..."

The price of crude oil also peaked that summer before plummeting 72% by December 2008. This drastic reduction in the price of crude oil was largely attributed to a decrease in global demand caused by the slowing global economy that characterized the final months of 2008. It also gave evidence of the real driver of commodity and consumer prices around the world. As long as we live in a petroleum economy, crude oil will determine not only how much we pay at the fuel pump but the price of virtually all the products and services that it touches.

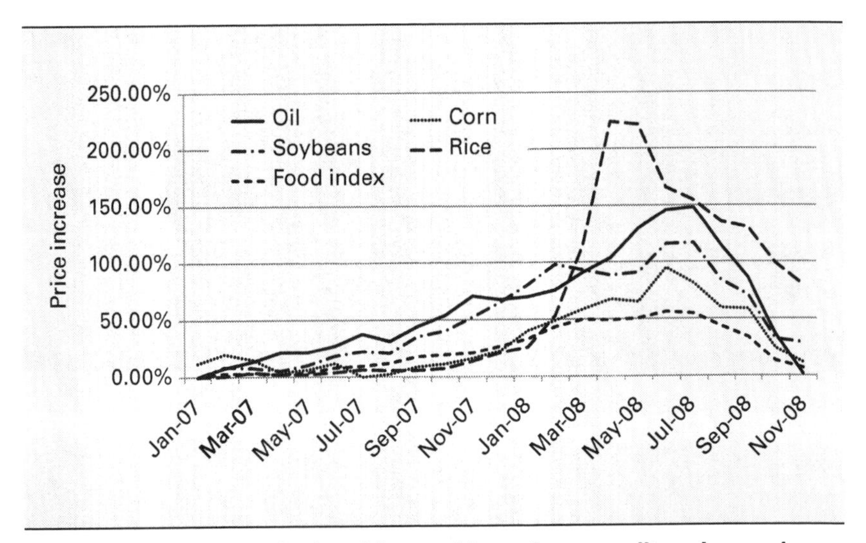

The price of petroleum is the ultimate driver of commodity prices as long as we live in a petroleum economy. *Source: International Monetary Fund*

Is biofuels agriculture environmentally sustainable?

Biofuels – particularly ethanol – have created concerns about their negative impact on the environment, a rather ironic situation given that biofuels are frequently heralded as the environmentally friendly alternative to fossil fuels such as crude oil. This concern recognizes that first-generation ethanol relies on traditional row crop agriculture, which in turn relies heavily on chemical fertilizers and pesticides to increase crop yields. These chemicals can find their way into waterways and groundwater, potentially contaminating drinking water supplies. While this has been a longstanding criticism of modern agriculture in general, the expansion of conventional row crop agriculture to meet increasing demand for crop-based ethanol inevitably increases the pollution associated with conventional crops and cropping systems.

Agriculture can have both positive and negative impacts on the environment depending upon how it is practiced. Early cultivation of cotton and tobacco in the southern United States was notorious for quickly depleting soil fertility, requiring frequent opening of new lands for production of these crops. Unsustainable agricultural practices result in soil erosion, water pollution, air pollution, reduced biodiversity, loss of soil fertility, and net positive emissions of greenhouse gases from the soil to the atmosphere.[32]

Soil fertility is a function of organic carbon and inorganic nutrient content of the soil. Organic carbon is a natural product from the decay of plant material in oxygen-poor environments. The amount of organic carbon in the soil increases or decreases from year to year depending on tillage practices. Conventional moldboard plowing and row cropping, by exposing buried carbon to erosion and oxidation, contribute to the loss of soil fertility. Modern conservation tillage practices and the substitution of perennial crops for row crops substantially reduce and can even reverse these losses.

The amount of inorganic nutrients (potassium, phosphorous, and nitrogen) in soils of natural ecosystems is determined by a balance among wind and water transport to and from the soil, uptake by standing biomass, and release by decaying biomass. Intensive agriculture disrupts this balance by removing inorganic nutrients with the harvested biomass, which must be periodically replaced by application of fertilizer. Corn, in particular, is notable for its demand for nitrogen fertilizer, which alone represents about one-half the energy input to rain-fed corn production. As subsequently described, nitrogen and phosphorous can contribute to water pollution through chemical run-off while the nitrogen can contribute to greenhouse gas emissions. In general, the establishment of dedicated energy crops in place of conventional row crops improves soil fertility. However, this may not be true if the dedicated energy crops are established on fallow lands or pasture, especially in the early years of establishment.

Soil erosion is the transport of soil from one location to another, either by wind or water. Rain and wind combine to erode topsoil from farmland at a rapid rate: as much as 10 times the rate of replenishment in the U.S. and 30-40 times the rate of replenishment in countries such as China and India.[34] In the process, organic carbon and inorganic nutrients lost from one location are gained by other locations; thus, erosion is not inherently bad. In practice, though, the net effect of soil erosion is to rob fertility from lands under intensive agriculture, which reduces crop productivity, and distributes soil and nutrients to unwanted locations, thus making them pollutants. As additional land is converted to farmland, the rate of erosion can be expected to increase. This is not a new problem – the Dust Bowl of the 1930s is an extreme example – but there are concerns that increased ethanol production will further accelerate this rate. While the use of conservation tillage can diminish erosion, it adds to the cost of crop production. Perennial crops have soil erosion rates that are one to two orders of magnitude less than for row crops such as corn and soybeans.

Water pollution arises from both water-induced soil erosion and the leaching of chemicals (inorganic nutrients, herbicides and insecticides) from soils.[35] Transport of soil from cultivated fields into waterways is responsible for the notoriously muddy rivers and streams of the Midwestern United States. Suspended solids can dramatically change aquatic ecosystems, driving out game fish, for example, which are replaced by bottom feeders. Soils washed into reservoirs and lakes drop out of suspension to form sediments that seriously reduce the water-holding capacity of these impoundments, affecting both animal and human activities. Suspensions of soil can also facilitate transport of agricultural chemicals, which can attach to individual soil particles.

Leaching of chemicals from soils is roughly proportional to their application rates to cropland. Although comprehensive data is not available, as little as 40% of nitrogen applied to an annual crop is incorporated into plant matter, the balance being volatilized or leached from the soil. Leaching of inorganic nutrients, herbicides, and insecticides can be expected to be considerably worse for row crops than perennial energy crops. Ammonia, the usual form of nitrogen in fertilizer, is oxidized to nitrates when exposed to air. Nitrates are readily leached from soils and can appear in both well water and river water at concentrations exceeding the 10-ppm human health standard set by the U.S. Environmental Protection Agency (EPA).

Phosphorous, which binds tightly to soil particles, is washed from fields as a result of soil erosion. Phosphorous represents a particular threat to aquatic ecosystems. In a process known as eutrophication, phosphorous promotes the rapid growth of algae near the surface of bodies of water. Eventually the algae die and their decomposition reduces dissolved oxygen to levels too low to support aquatic organisms – a general die-off occurs in the body of water. The problem can reappear periodically since phosphorous accumulates in the sediments of lakes and streams. Rapid run-off during a storm can stir up sediment and release phosphorous in another cycle of eutrophication.

Herbicides and insecticides are, by definition, toxic chemicals used to control plant and animal pests. Their environmental and health effects at low concentrations are not well known but they have been implicated as potential carcinogens and endocrine disrupters. They can be directly leached from the soil or transported by attachment to eroded soil particles. The lower pesticide application rates and the lower soil erosion rates associated with perennial energy crops suggest that pesticide pollution will be considerably less than for conventional row crops.

The contribution of crop production to air pollution is of two types: exhaust emissions from tractors and other production machinery and dust and gas arising from tilled soils. Exhaust emissions from internal combustion engines include nitrogen oxides, carbon monoxide, unburned hydrocarbons, and fine particulate. The total emissions from production machinery are small and diffuse compared to those from the transportation and utility sectors of the economy.

Several gaseous pollutants can arise from cultivation. Soil nitrogen, either from nitrogen-fixing bacteria or synthetic fertilizers, is converted to nitric oxide (NO) and nitrous oxide (N_2O) by microbial processes in wet, anaerobic soils. Methane, a greenhouse gas, is also produced by microbial processes in tilled soils; however, the vast majority of methane emissions come from animals and anaerobic digestion of animal wastes.

Nitric oxide released to the atmosphere is further oxidized to nitrogen dioxide (NO_2), which can produce both acid rain and promote ozone formation. Environmental degradation from acid rain includes acidification of lakes, which kills aquatic organisms, and damage to forests and crops. Ground level ozone is a health hazard that has proved a serious concern in large cities where NO is formed by automobiles and power plants. It is not known whether nitrogen oxide (NO_x) emissions from agricultural lands are a substantial problem.

Nitrous oxide is relatively stable to chemical reaction compared to NO, but it is a strong greenhouse gas. The background level of N_2O in the atmosphere has been increasing during historical times. Some researchers attribute this trend to intensive agriculture. The roles of N_2O and methane (CH_4) in global warming are discussed in a later section of this chapter.

Biodiversity describes an environment characterized by large numbers of different species of plants and animals. Agriculture traditionally has strived for just the opposite – a monoculture of one plant species where other plants and animals are considered pests. Production of dedicated energy crops has some prospects for improving biodiversity in the agricultural landscape. Certainly, the meadow-like setting of fields planted to perennial grasses and the forest-like setting of tree plantations more closely resemble natural ecosystems than do row crops. In fact, there are advantages in encouraging a certain degree of biodiversity in dedicated feedstock supply systems. Multi-species production systems could reduce the risks associated with pests. Adding nitrogen-fixing plants could reduce fertilizer applications. Other plant varieties might enhance erosion control.

Annual and perennial crops have different environmental impacts.

Cropping system	Soil erosion rate (Mg ha-1 yr-1)	N-P-K application rate (kg ha-1 yr-1)	Herbicide application rate (kg ha-1 yr-1)	Insecticide application rate (kg ha-1 yr-1)
Annual crops				
Corn	21.8	135-60-80	3.06	0.38
Soybeans	40.9	20-45-70	1.83	0.16
Perennial crops				
HEC	0.2	50-60-60	0.25	0.02
SRWC	2.0	60-15-15	0.39	0.01

Source: Reference 33

Can biofuels improve energy security?

Shifting reliance from imported petroleum to biofuels will not in itself improve U.S. energy security. Indeed, energy security is an issue for any country that relies on a single fuel source to meet the bulk of its energy needs. While many pundits and politicians have encouraged biofuel production within the U.S. as a means of making the country less dependent on foreign suppliers for its energy (the implication being that foreign oil producers can and will use oil embargoes as a geopolitical weapon much as OPEC did during the 1970s), dependency on domestic resources for energy poses its own set of problems. Survival in early agricultural societies depended on a bountiful harvest: crop disease, insect infestations, droughts and floods could all result in starvation during the winter and spring months. While famine is no longer a concern in the West thanks to improved transportation, high crop yields, and the development of food preservatives, the price of biofuels primarily depends on the price of the feedstock it is derived from which in turn depends on the productivity of the harvest (in addition to other factors such as demand). A country dependent on corn-based ethanol, for instance, would witness a rapid rise in transportation fuel prices (and a subsequent slowing of the economy) were a severe drought to strike the Midwestern farm states. While depending on a variety of biofuel feedstocks instead of just one and building up ample reserves of fuel would help mitigate this risk, any significant changes in growing seasons (such as those predicted by many global warming researchers) could cause severe fuel shortages in a biofuel-based economy that does not take these precautions.

Does it take more energy to produce biofuels than you get out of it?

The prominent biofuels critic David Pimentel is famously quoted as saying "It takes about 1.3 gallons of oil to produce one gallon of ethanol." The original attribution is difficult to trace, but it appears to have been an interview he gave to Minnesota Public Radio in 2005.[36] Whether he regrets this gratuitous and demonstrably false statement is not known but he has never retracted it. It is more difficult to track the origin of variations of this remark, including "it takes more energy to produce ethanol than you get out of it," which startles many people but happens to be true for all transportation fuels whether gasoline, ethanol, hydrogen, or electricity. The issue has become confused by a widespread misunderstanding of what is meant by energy balance in this debate over biofuels. Any treatise on biofuels would be incomplete without addressing this interesting yet, in some respects, irrelevant topic. In fact, there are many ways to define energy balance, depending upon what question is being asked about energy consumption. The question, for example, may be quite different for a plant engineer, the President's national security adviser, or an environmentalist.

For an engineer working at a biofuels plant, the focus is on the energy flows into and out of the processing facility. The relevant energy balance compares the useful energy output in the form of motor fuel to the energy inputs for the facility: electricity to run machinery, natural gas for drying and distillation, and the chemical energy contained within the corn to be processed. This is the classical definition of energy efficiency for conversion processes.

A different energy balance might be considered if the focus is on the amount of energy in the form of petroleum that is consumed in the production of motor fuel. This energy balance considers not only the processing facility but the landscape from which the biomass is obtained. The energy balance now becomes the ratio of motor fuel produced to the petroleum consumed in raising and harvesting the biomass as well as processing biomass into fuel. Since the United States is currently dependent upon imported petroleum from politically unstable regions of the world, this ratio might be termed "the national security advisor's energy balance." Clearly, this petroleum energy balance should be much larger than one, which it is for all commercially produced biofuels. Although the amount of fossil fuel employed in current grain ethanol production is not insignificant, very little of it is petroleum, imported or otherwise. As a result, the petroleum energy balance is at least 14 for grain ethanol compared to 0.81 for gasoline.[37] Unfortunately, some critics of biofuels have resorted to such inaccurate remarks as "it takes the equivalent of 1.29 gallons of gasoline to produce enough ethanol to replace one gallon of gasoline at the pump,"[38] which is equivalent to a petroleum energy balance of only 0.78. Whether an intentional effort to gain political advantage or merely a clumsy attempt to simplify complex issues, the net effect has been to confuse rather than edify the public and many policy makers.

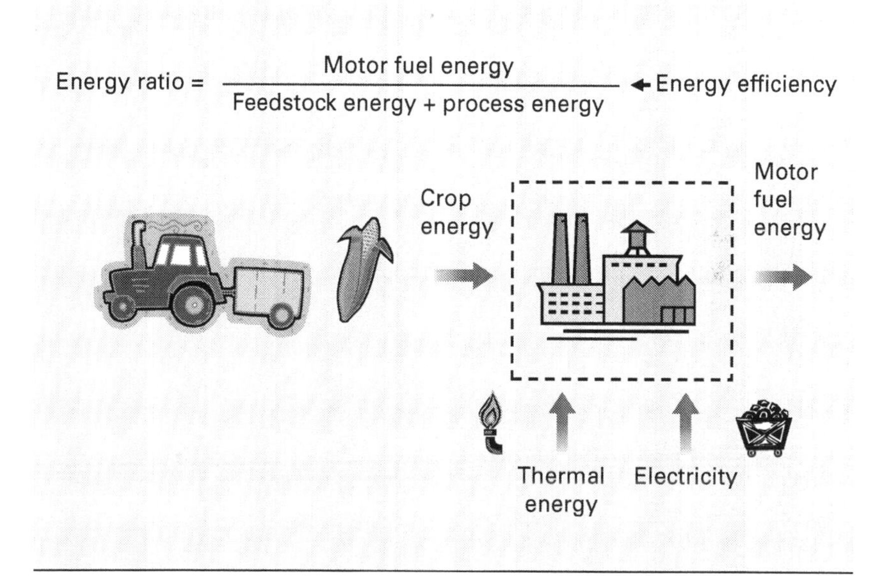

Biofuels energy balance from the perspective of a plant engineer. The ratio of motor fuel produced to energy consumed at the plant (including the energy content of the biomass) is called the energy efficiency.

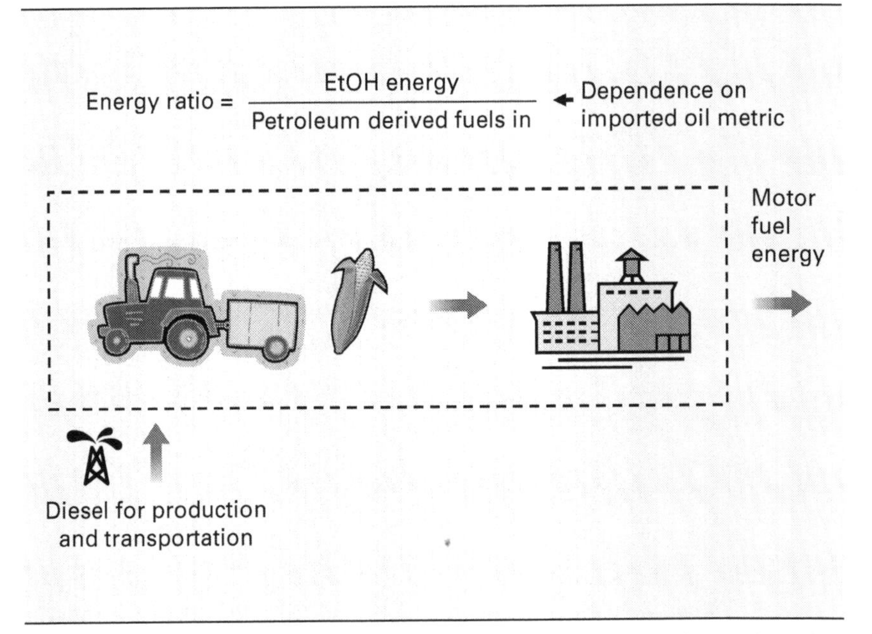

Biofuels energy balance from the perspective of the national security adviser. The ratio of motor fuel produced to the petroleum consumed in both producing and processing biomass is a measure of the fuels ability to help reduce dependence on imported petroleum.

However, most of the national debate on the energy balance of ethanol[39, 40] is about a third kind of energy balance that compares the output of motor fuel to the consumption of all forms of fossil fuels, including coal, natural gas, and petroleum. It considers energy consumption both in the growing and harvesting of biomass as well as its processing to motor fuel. This fossil energy balance is of particular interest to environmentalists since greenhouse gas emissions are closely correlated with the use of fossil fuels. It is an imperfect measure of environmental performance because fossil fuels show considerable variability in their greenhouse gas emissions per unit of energy released. Nevertheless, it is clear that the fossil energy balance should be greater than one if the goal is to reduce fossil fuel consumption in an effort to reduce greenhouse gas emissions. Several research groups have reported fossil energy balances for grain ethanol with values ranging from 0.44 to 2.1. Averaging over the values reported by 14 *different* study groups (to avoid replicating values of the same study groups reported in different publications) yields an energy ratio of 1.3.[40,41] In comparison, and there is little disagreement on this point, the environmental energy ratio for the production of gasoline from petroleum is only 0.81.

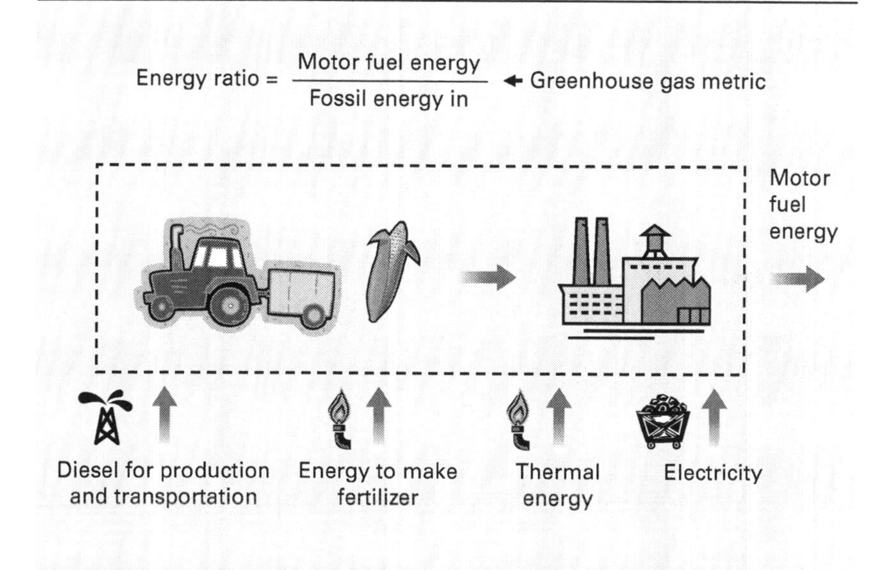

Biofuels energy balance from the perspective of an environmentalist. The ratio of motor fuel produced to the fossil fuel consumed in growing, harvesting, and processing biomass is a useful way to gauge the amount of greenhouse gas emissions associated with producing a particular biofuel.

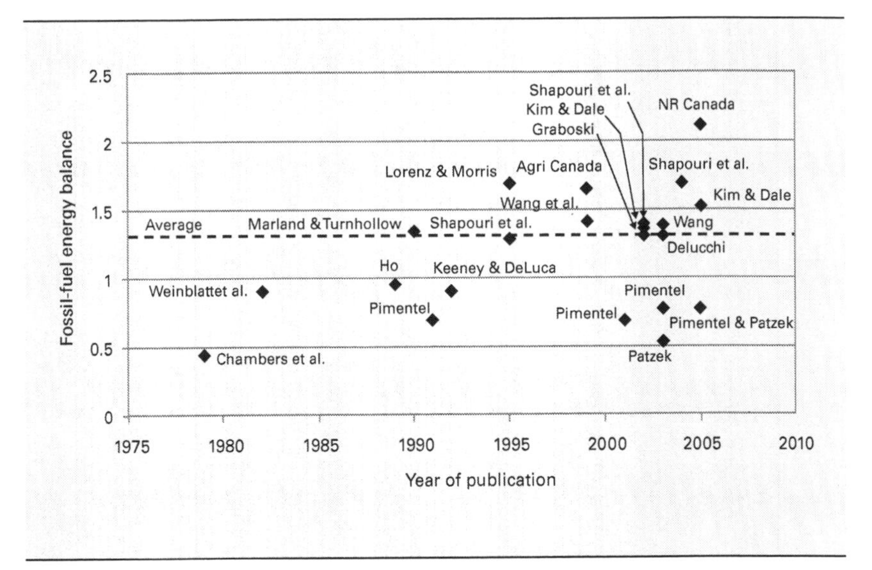

Researchers disagree about the fossil energy balance of grain ethanol. Although the ratio of motor fuels produced to fossil fuels consumed appears to be slightly greater than one, there is no reason the ratio could not be considerably higher. *Adapted from Reference 41.*

There are several reasons for the wide range of reported values for the fossil energy ratio in the production of ethanol. First, different study groups make different assumptions about the production yield of corn grain. For example, one study group averages corn yields over all fifty states with the intention that this best represents a national average for corn yield while another averages yields over the top ten corn producing states, arguing that these are where grain ethanol plants are clustered. Second, there are major disagreements about the amount of energy needed to produce anhydrous ammonia fertilizer. Third, there is no consensus on the amount of ethanol that can be produced from a bushel of grain, probably because this number depends on the age and size of the fermentation facility. Finally, there are questions as to the amount of fossil energy consumed within the production facility. Clearly, a large amount of natural gas is consumed in drying DDGS and distilling ethanol, but it is difficult to accurately assess energy consumption in an industry that is rapidly growing and changing. Ultimately, the disagreements among researchers likely reflect the difficulty of assigning average values for these parameters to the whole industry. Very likely older and smaller corn ethanol plants operate with fossil energy balances less than unity while larger, more modern facilities operate above unity.

Nevertheless, there is clearly room for substantial improvement. A fossil energy ratio of 1.3 means that the ethanol product contains only 30% more energy than the

Fossil-fuel energy balances range widely for different kinds of biofuels

Fuel (feedstock)	Fossil energy balance*
Cellulosic ethanol	2-36
Biodiesel (palm oil)	~9
Ethanol (sugar cane)	~8
Biodiesel (waste vegetable oil)	5-6
Biodiesel (soybeans)	~3
Biodiesel (rapeseed, EU)	~2.5
Ethanol (wheat, sugar beets)	~2
Ethanol (corn)	~1.5
Diesel (crude oil)	0.8-0.9
Gasoline (crude oil)	~0.8
Gasoline (tar sands)	~0.75

*Energy contained in the listed fuel per unit of fossil fuel input including biomass production and processing.

Source: Reference 42.

fossil fuel employed in the overall process of growing the grain and processing it into motor fuel. Many kinds of current and future biofuels have fossil energy balances that are several times higher than for the grain ethanol industry. For example, sugar cane ethanol has a fossil energy balance of approximately eight, arising primarily from the use of processing residues for heat and power within the ethanol plant. Several things could be done to reduce the use of fossil fuels in ethanol production: tractors could run on pure biodiesel; corn stover could be the energy source for fertilizer production; wood residues could be used as the source of energy for drying and distillation or less energy intensive operations could be employed. Reductions in fossil fuel use would increase the fossil energy ratio: values of ten or higher could readily be achieved. Advanced biorefineries are expected to contribute to these improvements.

Could it be that biofuels emit more greenhouse gases than petroleum-derived fuels?

The debate over the energy balances described in the previous section is in many respects a distraction from a more important question about renewable fuels: can they significantly reduce greenhouse gas emissions compared to fossil fuels? The original position of many scientists and environmentalists was that renewable fuels would dramatically reduce greenhouse gas emissions. Carbon dioxide (CO_2) from burning gasoline and other fossil fuels ultimately derives from carbon-rich molecules extracted from deep underground. This movement of carbon from geological deposits to the atmosphere contributes to greenhouse gas warming in an essentially irreversible manner in the time scales over which the fossil resources are consumed. In contrast, CO_2

from burning biofuels derives from carbon-rich plant molecules that were recently synthesized by living biomass from water in the hydrosphere and CO_2 in the atmosphere. Thus, photosynthesis continuously cycles carbon between the atmosphere and growing biomass, leading to the notion that biofuels are "carbon neutral."

In recent years this concept has been viewed as an oversimplification of the contribution of biofuels to the Earth's carbon balance. Not only is CO_2 emitted during the burning of fuel, it and other greenhouse gases are emitted during agricultural production of biomass and processing of it into biofuels. Natural ecosystems store carbon in standing biomass (trees and grasses) and as soil carbon. Depending upon the kind of ecosystem and the tillage system introduced, this conversion to farmland can release significant quantities of CO_2 to the environment. Obviously, a mature forest contains a considerable amount of carbon and burning it will produce a massive release of CO_2 to the atmosphere. Grasslands store a relatively small amount of carbon in standing biomass, but often stores large amounts of carbon in the soil. Tilling the soil can expose some of this buried carbon to the atmosphere, where it is oxidized to CO_2. Diesel-powered tractors and other equipment used for plowing, planting, fertilizing, and harvesting also emit CO_2. Nitrogen fertilizer consumes significant fossil fuels in its production, which adds to the greenhouse gas burden of agriculture. Once applied to the soil, microbial oxidation of nitrogen fertilizers can release N_2O, a potent greenhouse gas, to the atmosphere.[43] Transport of corn to ethanol plants and distribution of the ethanol fuel to consumers rarely is accomplished with ethanol-fueled vehicles, and these CO_2 emissions must be counted against ethanol. The extensive drying and distilling processes in an ethanol plant are usually accomplished with natural gas or coal instead of biomass, despite the fact that this is the single largest source of greenhouse gases in the production of ethanol.

It is within the power of the grain ethanol industry to reduce greenhouse emissions from both biofuels agriculture and biofuels production. Tractors can be run on biodiesel instead of diesel fuel. Nitrogen fertilizer can be manufactured from renewable resources instead of fossil fuels. Tillage practices can be adopted that reduce soil carbon loss and N_2O emissions. Ethanol manufacture can be made more efficient. Biomass can replace natural gas and coal for process heat in ethanol plants. Dramatic improvements are possible because, unlike fossil fuels, biomass is not

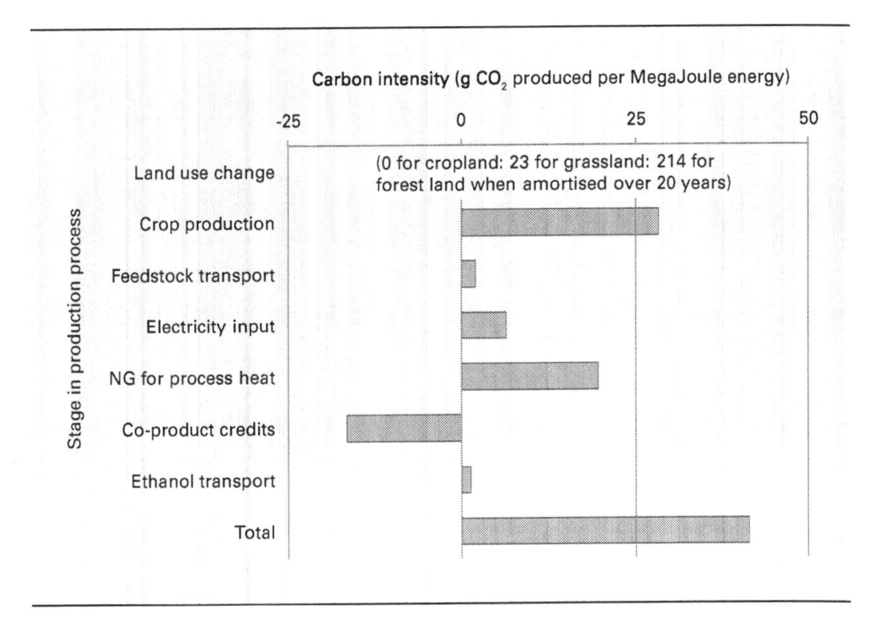

Biofuels production is not carbon neutral. This example demonstrates how fossil fuels find their way into all kinds of economic activities. Land use change shown here are direct impacts. *Source: Reference 44.*

inherently a net source of greenhouse gases. All that is required are appropriate incentives that encourage adoption of low-carbon agricultural practices and discourage reliance on fossil fuels. Unfortunately, little progress will be made while the biofuels industry is distracted by the efforts of some environmental groups to hold biofuels agriculture responsible for the greenhouse gas emissions arising from unsustainable food and feed production in other parts of the world - the so-called "indirect land use change" (ILUC) contribution to greenhouse gas emissions.

The biofuels boom that developed in the first decade of the twenty-first century manifested itself across the world in quite different ways. In the United States, processing plants were built to convert corn surpluses into ethanol. Brazil supplied its expanding ethanol industry with sugar cane grown principally on abandoned grazing lands.[45] In Southeast Asia, biodiesel was produced from palm oil grown on plantations that were established by clearing rainforests. These are all examples of direct land use changes to support biofuels production. Whether these changes increase or decrease greenhouse gas emissions depend upon the previous use of the land and the kind of tillage practiced. Diverting corn to ethanol production instead of livestock production does not directly affect greenhouse gas emissions from this land use change. On the other hand, burning down a rainforest to establish an oil palm plantation will temporarily boost CO_2 emissions until the plantings become a mature stand of trees. Environmentalists, alarmed at the rate of deforestation in the tropics, raised concerns about the role of biofuels in this destruction. Their protestations were largely dismissed in the developing world where economic growth trumped ecosystem preservation.

The voice of the environmental movement is more

sympathetically received in the developed countries of the world where affluence affords people more time and money to develop an environmental ethic that values the preservation of rainforests for non-economic purposes. Out of frustration to stop direct land use change in the developing world, the concept of indirect land use change was born with the goal of enlisting the support of the developed countries of the world to halt deforestation in the tropics.

Indirect land use change (ILUC) occurs when an acre of cropland diverted from food production to biofuels production forces the conversion of an acre of forestland or grassland into farmland to avoid a deficit in food supply. If this new farmland comes from burning rainforests, it is easy to calculate that the CO_2 emissions from the fires would exceed for many years the emission reductions achieved by displacing gasoline with renewable fuel. If this new farmland comes from plowing grasslands, then CO_2 and N_2O emissions from the soil will increase, adding to the greenhouse gas burden of the atmosphere. Environmentalists argue that the biofuels industry should be made responsible for these emissions even though it does not directly use this new cropland since purportedly the land would not have been converted to agriculture but for their diversion of corn and soybeans to biofuel production. Rather than assigning this greenhouse gas burden to the crops grown directly on the newly created farmland, they are assigned to biofuels produced from crops in the United States and other developed countries.

Timothy Searchinger, a former senior attorney at the Environmental Defense Fund, and Joe Fargione, a researcher at the Nature Conservancy, and their collaborators were among the first to quantify the potential impact of ILUC on greenhouse gas emissions. In early 2008 both groups of researchers reported that corn-based ethanol produces higher net greenhouse gas emissions per unit of energy delivered than petroleum-based gasoline.[46, 47] Because the U.S. Congress and California Legislature had recently passed energy bills requiring alternative fuels to perform substantially better than gasoline in terms of greenhouse gas emissions, the findings of Searchinger and Fargione raised the ironic prospect that renewable fuels could be banned and gasoline production expanded to meet future U.S. needs for transportation fuels.

Starting in 2011, California's low carbon fuel standard (LCFS) requires all gasoline or alternative fuels substituted for gasoline to release less than 95.6 grams of CO_2 equivalent greenhouse gas emissions per megajoule of energy content of the fuel (g CO_2 per MJ) on a life cycle basis.[48] After 2020, the LCFS decreases to 86.3 g CO_2 per MJ (diesel fuel and diesel fuel substitutes are slightly lower). These greenhouse gas reductions include not only tailpipe emissions but also emissions associated with the production and distribution of the fuel. In the case of biofuels, this includes both direct land use impacts and indirect land use change impacts. Searchinger's analysis assigned to corn ethanol direct GHG emissions of 74 g CO_2 per MJ, which would meet CARB's low carbon fuel standard. However, inclusion of ILUC as calculated by Searchinger increased corn ethanol GHG emissions to 177 g CO_2 per MJ, a 105% increase over gasoline.

The U.S. EPA uses a slightly different approach because the language of the 2007 Energy Independence and Security Act specifies "life cycle GHG thresholds" for different fuels as percent reductions compared to a gasoline baseline.[49] The EPA's gasoline baseline (2005) is 93 g CO_2 per MJ. The threshold for renewable fuels (corn ethanol or biodiesel) produced in new plants is a 20% reduction over the gasoline baseline (existing plants receive a waiver from this requirement) while cellulosic biofuels must achieve at least a 60% reduction in life cycle greenhouse gas emissions compared to the gasoline baseline. Thus, corn ethanol must emit no more than 74 g CO_2 per MJ while cellulosic ethanol is limited to 37 g CO_2 per MJ.

Lifecycle Greenhouse Gas Thresholds for Biofuels as Specified by EISA

Renewable fuel	20%
Advanced biofuel	50%
Biomass-derived diesel	50%
Cellulosic biofuel	60%

Percent reduction relative to 2005 gasoline baseline.

Source: Reference 49.

In its first attempt at life analysis of greenhouse gas emissions for biofuels in 2009, the EPA found that corn ethanol, sugar cane ethanol, and cellulosic ethanol all met the thresholds for reducing GHG emissions mandated by EISA.[49] But the EPA also determined that indirect land use change impacts should be included in these determinations, which can dramatically influence the outcome depending upon the assumptions employed. Among its most pessimistic scenarios, a thirty year "payback period" was assumed for the ILUC contribution to emissions. This scenario resulted in corn ethanol being disqualified as an acceptable renewable fuel unless the ethanol plant was operated in conjunction with a combined heat and power (CHP) plant, which is rarely practiced in the U.S. Sugar cane and cellulosic ethanol were more fortunate, qualifying as renewable fuel even when ILUC was included. Much to the chagrin of some environmental groups, the EPA also contemplated a more forgiving payback period of 100 years, which would have qualified grain ethanol under the 20% reduction threshold. Everyone should be chagrined at such arbitrary rule setting, which is based on assumptions that in some cases are ultimately unknowable because they depend upon unpredictable human behavior.

In 2009 Professor Dermot Hayes at Iowa State University, one of Searchinger's co-authors on the 2008 Science

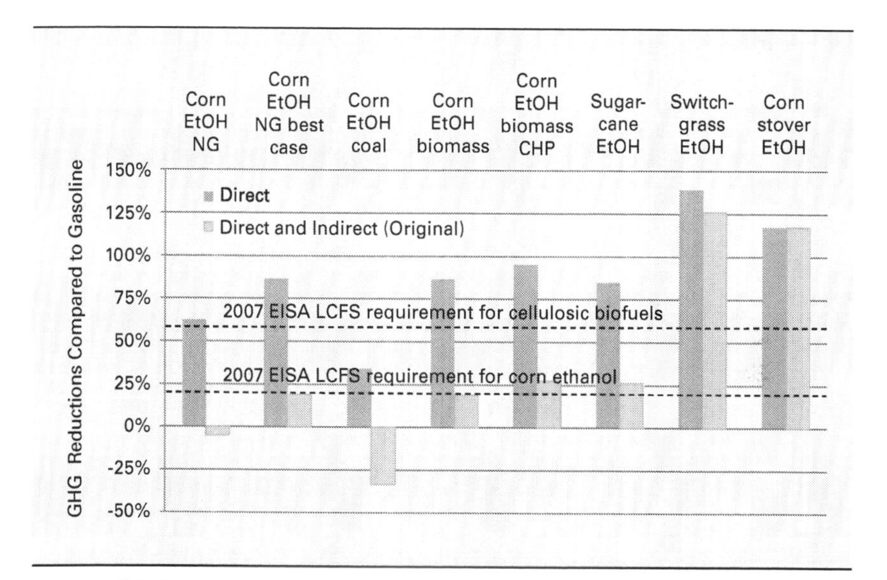

In 2009 ILUC appeared to be game changing. Inclusion of indirect land use change in calculating greenhouse gas emissions as originally proposed would have disqualified biofuels if produced in corn ethanol plants. Corn EtOH – corn ethanol plant; NG – natural gas for process heat; Coal – coal for process heat; Biomass – biomass for process heat; CHP – combined heat and power. *Source: Reference 49.*

paper that brought so much public attention to ILUC, re-evaluated the assumptions about agricultural production, land-use change, and greenhouse gas (GHG) emissions.[50] The original study estimated that it would take 167 years before the substitution of corn ethanol for gasoline reduced direct emissions of GHG sufficiently to account for indirect GHG emissions from ethanol-induced conversion of natural ecosystems to agriculture. Using the same methodologies as the original study, Hayes and his colleagues were able to replicate the original "payback" period of 167 years. However, they discovered that this payback period was exquisitely sensitive to the assumptions employed in the analysis. For example, simply enhancing corn yields by 1% over 10 years reduced the payback period by 136 years. Several of the alternative assumptions explored by Hayes and co-workers in combination could reasonably reduce the payback period to zero. The authors concluded that: "The lesson for policymakers is that results from economic models depend heavily on assumptions, and because we are trying to predict long-run human behavior, there can be legitimate differences in these assumptions."

The theory of ILUC impacts received wide circulation at a time when the argument seemed eminently reasonable: corn prices were headed toward record-high levels, rice and wheat were being hoarded by some countries, food prices were increasing, in some developing countries people were rioting over access to food, and rainforests were reported to be disappearing at a furious rate. Although lacking hard evidence, many commentators had little trouble condemning the biofuels industry for many if not all of these developments.

But the case for ILUC is flawed in several important respects. First, increases in corn supply have to date outstripped the incremental increase in demand by the ethanol industry – there is no corn deficit driving the expansion of agriculture around the world. Agriculture is expanding because we have 800 million more mouths to feed than we did 10 years ago and many of them aspire to the high protein diets of the West. As explained earlier in this chapter, the energy efficiency of converting corn to beef is so low that it will take twenty times as much land to feed the growing population hamburgers as it does to feed them corn flakes. In this respect, the problem is more a matter of "food vs. feed" than "food vs. fuel." Undoubtedly, the sudden inflation in the price of corn between 2007 and 2008 encouraged farmers to redouble their efforts to grow corn, but we have also seen that the price of corn does not correlate with the amount of ethanol in production around the world.

Changes in the price of corn, other commodity crops, processed foods, and even ethanol are driven primarily by the price of petroleum. One could even reasonably argue that since petroleum prices drive corn prices and corn prices encourage or discourage the expansion of agriculture, it is ultimately petroleum that should be held responsible for destruction of rainforests and grasslands and the greenhouse gas emissions entailed. But the world is more complicated than that. As pointed out by Keith Kline and Virginia Dale of Oak Ridge National Laboratory:[51]

> "...field research...consistently finds that land-use change and associated carbon emissions are driven by interactions among cultural, technological, biophysical, political, economic, and demographic forces within a spatial and temporal context rather than by a single crop market."

The second flaw is the difficulty in establishing how people in the developing world respond to changes in world agricultural markets. As corn prices increased around the world, a number of responses were possible, none as dire as burning down rainforests. Other possibilities include: consumers in developed countries reduced their consumption of animal protein; corn distributors took care to avoid spoilage and waste; subsistence farmers learned to grow cash crops; Afghan farmers grew more wheat and less poppies; fallow farmland and pasture land were planted to food crops; farmers in the developing world embraced modern agriculture to increase yields. Like the suggestion that rainforests were burned down in the name of biofuels agriculture, these are merely conjectures until they are formulated into testable hypotheses. Certainly none of them should be used to formulate public policy until strong experimental evidence is unable to dismiss them as conjecture.

Third, empirical evidence does not support either direct land-use change (DLUC) or ILUC as an explanation for the disappearance of rainforests. If sugar cane production on abandoned cropland and pastureland in Brazil has pushed agriculture and grazing into the Amazonian rainforest, then there should be a correlation between the rate of Brazilian rainforest destruction and the increase in sugar cane ethanol production. The data from 1989 to 2009 does not support this hypothesis. If it was as simple a matter as the price of soybeans encouraging the conversion of rainforests to agriculture, then we should find a correlation between these two factors. The data from 1989 to 2009 does not support this hypothesis. A key assumption behind the ILUC argument is that rising food prices are directly responsible for pressure on the remaining rainforests of the world. If this was true, then we would expect a correlation between the rate of Brazilian rainforest destruction and the commodity food price index. The data from 1989 to 2009 does not support this hypothesis. While a lack of available data precludes a similar analysis of global deforestation rates, the Food and Agricultural Organization of the United Nations (FAO) reported in 2010 that global deforestation actually fell by 19% from the last decade of the 20th century to the first decade of the 21st century, despite the rapid growth in biofuel production during the last decade.[52]

Fourth, the inclusion of ILUC in the calculation of the LCFS is unlikely to substantially reduce the pace of deforestation in the tropics. If the ILUC argument is correct,

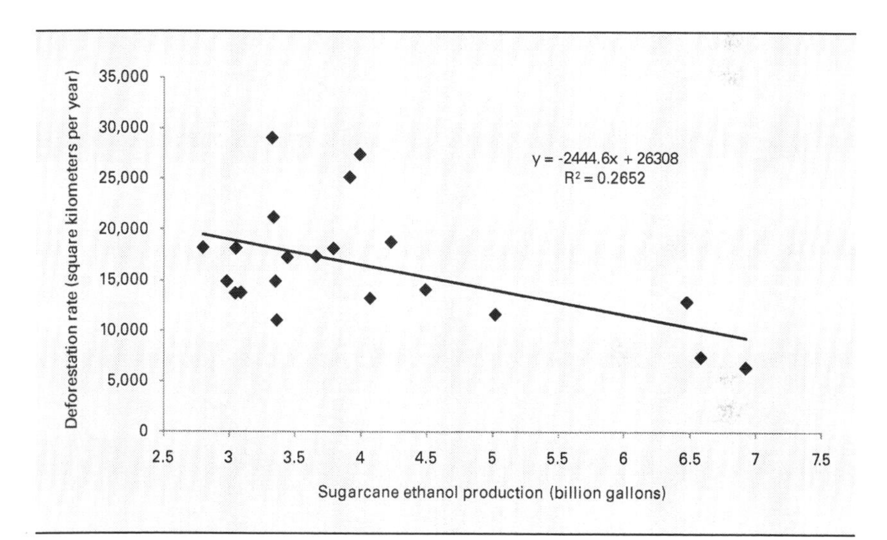

There is no positive correlation between the amount of sugar cane ethanol produced in Brazil and the deforestation rate in the Amazonian rainforest of Brazil. The solid line is a "best fit" of the data, which shows a negative linear correlation. However, the regression coefficient (R^2) of 0.2652 indicates that the statistical significance is small. (a value of 1.00 is a perfect correlation). *Source: United Nations Food and Agriculture Organization (deforestation data) and Renewable Fuels Association (ethanol production data).*

biofuels agriculture is responsible for less than 25 million acres of the almost 500 million acres of rainforest that have disappeared in the last decade. Something else is responsible for the epidemic of deforestation. The answer is likely to be found in unsustainable agricultural practices in the tropics, which is still premised in many respects on primitive "slash and burn" farming. Lands in the tropics store most of its carbon in standing biomass since microbial activity consumes soil carbon almost as fast as it is formed from decaying biomass. Thus, the soils are thin and quickly worn out by conventional tillage of row crops like corn and

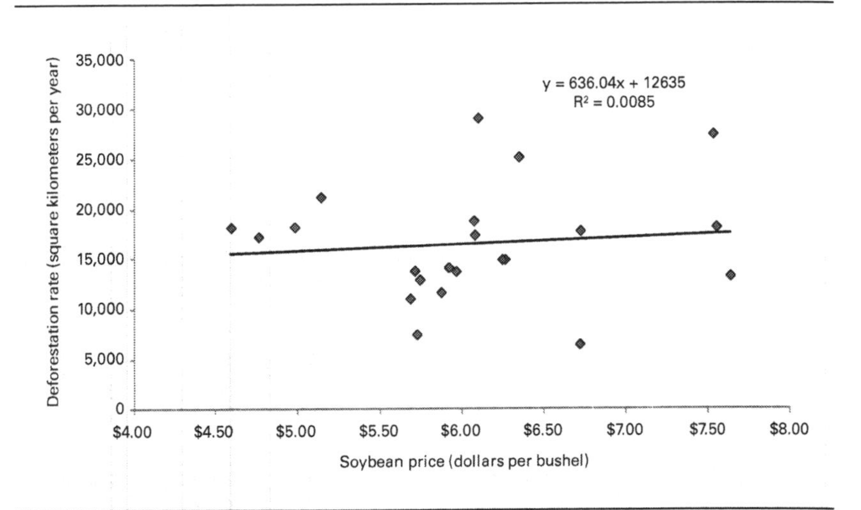

There is no correlation between the price of soybeans and the deforestation rate in the Amazonian rainforest of Brazil. The solid line is a "best fit" of the data, which shows a small positive linear correlation. However, the regression coefficient (R^2) of 0.0085 indicates virtually no statistical significance (a value of 1.00 is a perfect correlation). *Source: United Nations Food and Agriculture Organization (deforestation data) and Index Mundi (soybean price data).*

soybeans. When this occurs, tropical farmers seek out new forests and grasslands for conversion to agriculture.

Although this kind of "land churning" results in destruction of virgin forests and grasslands, it also is accompanied by the emergence of restored natural ecosystems. Whereas the annual rate of deforestation worldwide is about 38 million acres, the United Nations has estimated that over 2 billion acres of replacement forest is growing in the tropics, which includes former farmland and degraded forests.[53] This reversion to forestlands is not surprising in places like southeastern United States, but it is also happening in

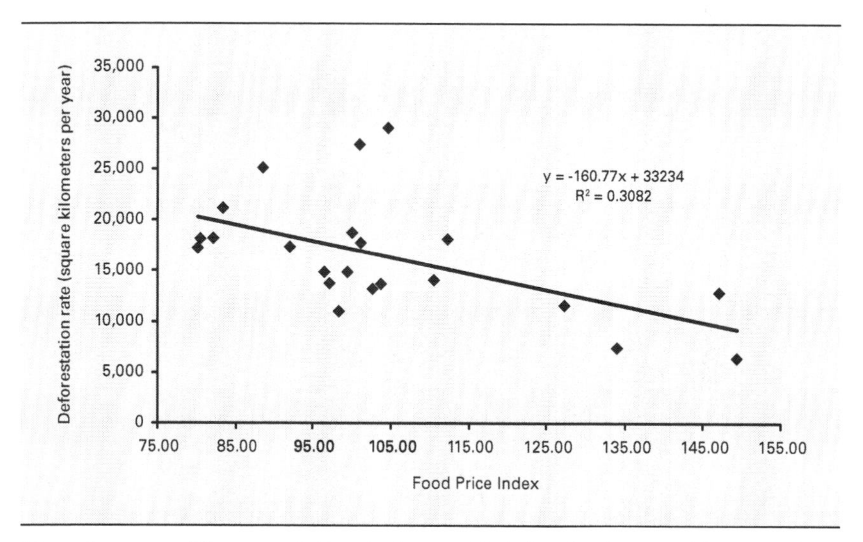

There is no positive correlation between high food prices and increased deforestation of the Amazonian rainforest of Brazil. The solid line is a "best fit" of the data, which shows a small negative linear correlation. However, the regression coefficient (R^2) of 0.3082 indicates only a weak statistical significance correlation (a value of 1.00 is a perfect correlation). *Source: United Nations Food and Agriculture Organization (deforestation data) and Index Mundi (food price index).*

tropical countries such as Panama, where rainforests are being destroyed at a rate of 1.3 percent each year but the pace of secondary forest emergence is increasing at 4 per cent every year.[54] Ultimately, these replacement forests will sequester large amounts of carbon, but only to the extent that the destruction of the original forests added greenhouse gases to the atmosphere. It also raises serious concerns about loss of habitat and replacement by ecosystems that differ in some respects from the original ones. If agriculture is responsible for a significant portion of the world's greenhouse gas emissions, then it seems short sighted not to expect both food and fuel agriculture to participate in efforts to control global climate change.

Early in 2010, the EPA announced revisions to statutory requirements to meet the Renewable Fuel Standard program outlined in the Energy Independence and Security Act of 2007. In addition to reducing near-term targets for cellulosic biofuels, which markets were clearly not able to meet, the revisions included determinations that modern corn ethanol plants fired with natural gas meet the mandated 20% greenhouse gas reduction; biodiesel plants meet the mandated 50% reduction; and cellulosic ethanol plants comply with the mandated 60% reduction.[55] What changed in a year's time? As explained by the EPA, "as the state of scientific knowledge continues to evolve in this area, the life cycle GHG assessments for a variety of fuel pathways are likely to be updated."[56] These updates included new studies that showed crop yields to increase with higher crop prices, which reduces demand for new land; new research that shows distillers' dried grains to be an efficient animal feed, meaning corn demand and exports as a result of biofuels agriculture are not as impacted as originally assumed; and new pasturelands are likely to be established from existing grasslands rather than from destruction of forestland. In other words, the assumptions have changed.

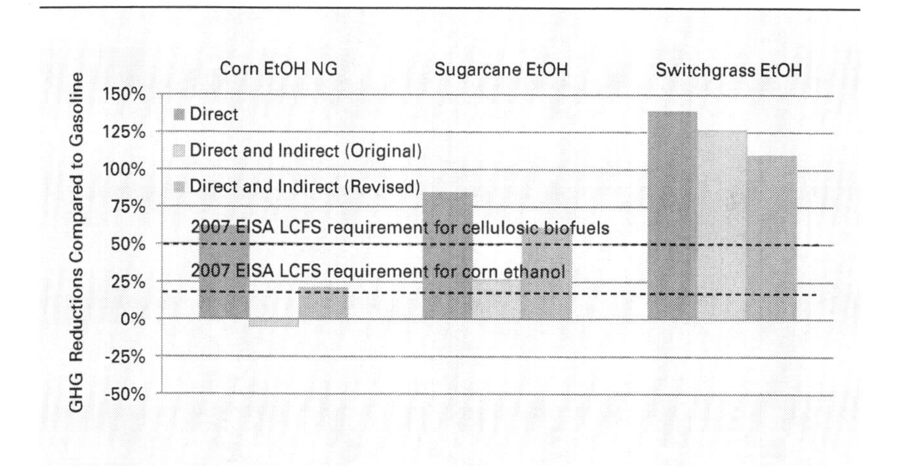

Updating ILUC assumptions. In 2010 the EPA revised its ILUC calculations, resulting in corn ethanol meeting the federal LCFS requirement for this technology. Corn EtOH NG – corn ethanol plant using natural gas for process heat; sugar cane EtOH – sugar cane ethanol plant using bagasse for process heat; switchgrass EtOH – cellulosic ethanol plant using lignin for process heat. *Source: Reference 55.*

8

How can we turn lipids into advanced biofuels?

The nature of lipid feedstocks

Lipids are a large group of hydrophobic, fat-soluble compounds produced by plants and animals for high-density energy storage. These include fats, sterols, triglycerides and waxes. Triglycerides, commonly known as vegetable oils, are among the most familiar form of lipids and have been widely used in recent years for the production of biodiesel. Until recently, the potential for significant market penetration of lipid-based biofuels was considered small because of the low productivity of traditional lipid-rich crops like soybean and rapeseed, which yield only 50 - 130 gallons of biodiesel per acre.[1] Sunflower, one of the most highly productive vegetable oil crops, only produces 60 – 176 gal/acre compared to 620 – 930 gallons of ethanol per acre of corn crop. Even accounting for the lower energy content of ethanol, the fuel energy obtained per acre is three to seven times higher for corn ethanol than oilseed biodiesel and even higher for cellulosic biofuels. For this reason little attention was given to further developing lipid-based fuels until recently.

The drivers for renewed attention to lipid-based fuels are two-fold. First, lipids are highly reduced compounds, containing very little oxygen, and in some respects resemble long-chained hydrocarbons found in gasoline. As will be subsequently described, lipids can be upgraded in a fashion similar to hydroprocessing of petroleum to yield hydrocarbons that are essentially indistinguishable from gasoline, diesel fuel, or aviation fuel. Both the U.S. military[2] and the aviation industry[3] are keenly interested in so-called "drop-in" biofuels that can directly substitute for their petroleum fuels without modification of their current infrastructure or operating procedures. They have no use for ethanol, whose hydroscopic properties make it unsuitable as aviation fuel. Hydrocarbons derived from lipid biomass, on the other hand, are very attractive as a low carbon aviation fuel.

The second driver for renewed interest in lipid-based fuels is the prospect for growing either very high yielding

Lipid based fuels come in first and last place when ranked for fuel productivity. Among biofuels options, biodiesel from oilseed crops like sunflowers yields the fewest gallons per acre while biodiesel from microalgae yields the most.

Process	Feedstock	Theoretical fuel yield (gallons per ton)	Biomass productivity (tons per acre)	Fuel productivity (gallons per acre)
Biodiesel	Sunflower	101-126	0.6-1.4	60-176
Grain ethanol	Corn grain	124	5-7.5	620-930
Biochemical cellulosic ethanol	Wood or grass	100-113	5-20	500-2260
Thermochemical cellulosic ethanol	Wood or grass	130-160	5-20	650-3200
Biodiesel	Microalgae	36-100	29-55	1044-5500

Source: Author

lipid feedstocks or feedstocks that thrive in environments not otherwise suitable for growing food crops. A number of drought- and salt-tolerant plants have been identified that produce lipids. This discussion will be limited to some of the more promising varieties: palm oil, jatropha, salicornia, and microalgae.

The Southeast Asian version of soybean oil, palm oil (*Elaeis oleifera*), is an edible vegetable oil derived from the oil palm tree that is primarily grown in Malaysia and Indonesia, although equatorial countries in South America and Africa are capable of growing it as well. Its popularity as a

Elaeis oleifera

crop has grown in recent years and as of 2004 was second only to soybean oil as the most widely-produced edible vegetable oil,[4] an increase driven in part by its use as a trans fat-free ingredient in the snack industry.[5] The similarities between oil palms and soybeans do not end there, however, as its use as a biofuel feedstock has generated much controversy. In addition to being caught in the middle of the food v. fuel debate thanks to its edible nature, its ability to thrive in tropical countries has caused some environmental groups to claim that it contributes to direct land use change and rainforest destruction as rainforests are cleared to make room for oil palm plantations.[6] Finally, increased demand for oil palm in Southeast Asia has periodically driven up its price, causing local unrest among populations that depend on the vegetable oil for cooking.

Imposing as these concerns are, it is unlikely that palm oil's popularity as a biofuel will diminish in the near future.

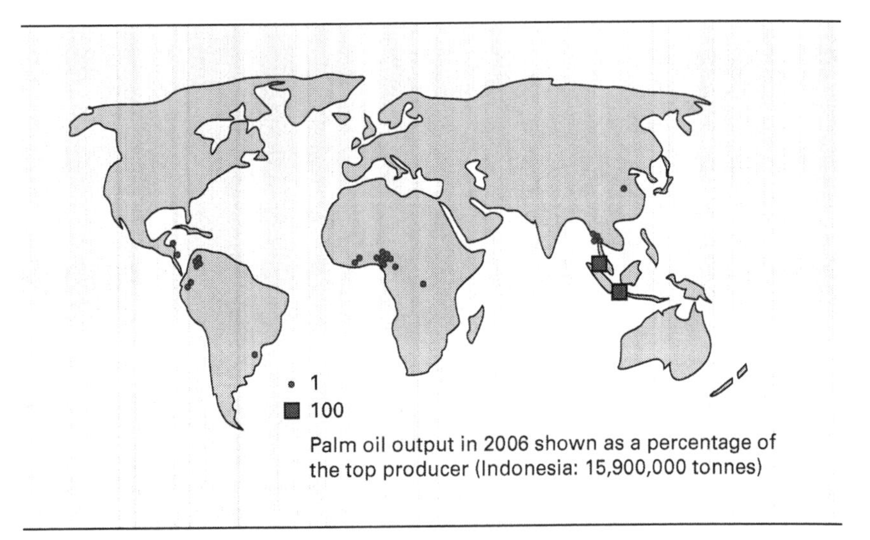

Palm oil plantations around the world. *Source: Adapted from Reference 9.*

This is largely because of the feedstock's substantial yield of more than 600 gallons per acre, which is nearly 10 times that of soybeans in terms of both energy yield and biofuel yield.[7] This makes it one of the highest-yielding lipid-based feedstocks currently being grown. The oil palm infrastructure is also well developed, palm oil having been used in both food and cosmetics for several decades before it became a popular biodiesel feedstock. Finally, the oil palm's overall yield is expected to increase as advances in cellulosic ethanol allow many of the byproducts created during palm oil processing to be used for production of additional biofuels,[8] an innovation that would have the added benefit of further reducing the feedstock's life cycle GHG emissions. However, environmentalists concerns that rainforest is being cleared to make way for oil palm plantations and its edible nature puts it in the middle of the food vs. fuel debate.

Jatropha is a genus of hardy bushes and trees originating in the Caribbean and now spread throughout the tropics that produces seeds containing up to 40% triglycerides. Yields for this inedible oil are estimated to be 200 – 400 gal/acre. The plant is not domesticated and it remains to be shown that it can be adapted to biofuels agriculture. Jatropha's use as a biofuel feedstock has received considerable media attention following the successful flight of a Boeing 747 jumbo jet powered partly by jatropha oil in late December 2008.[10] Jatropha oil, which is produced from the seeds of the jatropha plant, has been touted as a solution to the food v. fuel debate thanks to its ability to grow on marginal and non-arable land in harsh environments – the plant even has the ability to be intercropped with various fruits and vegetables. It is also an efficient biofuel feedstock, with potential yields far surpassing those of soybeans and corn.[11] The combination of these factors caused analysts at

jatropha curcas

Goldman Sachs to pronounce jatropha as "one of the best candidates for future biodiesel production."[12]

Analysts' optimism aside, jatropha faces several significant shortcomings that must be overcome before it can be considered a viable biofuel feedstock. The first of these is the fact that the plant has yet to be domesticated, resulting in a "crop" with yields that are anything but dependable. One industry observer noted that yield estimates range from as little as 0.5 ton to as much as 15 tons per hectare.[13] While its ability to grow under poor agricultural conditions is frequently touted, this is erroneously confused with the ability to thrive under such conditions. Additionally, its long-term impact on soil and the environment has yet to be studied. This has also hindered the ability to make accurate estimates of jatropha oil's production costs. Whereas crops like corn and soybeans have been heavily bred and genetically engineered to develop highly-efficient strains, genetic improvement of jatropha to improve its fuel yield is only in the very early stages of research.[14]

Long used for glassmaking and soapmaking, salicornia is an edible, salt-tolerant plant that grows in salt marshes and on beaches. Its seeds contain high levels of unsaturated oil suitable for biodiesel production.[15] Its main advantage over other lipid feedstocks does not lie in its oil content, however. As its name implies, salicornia has an extremely high salt

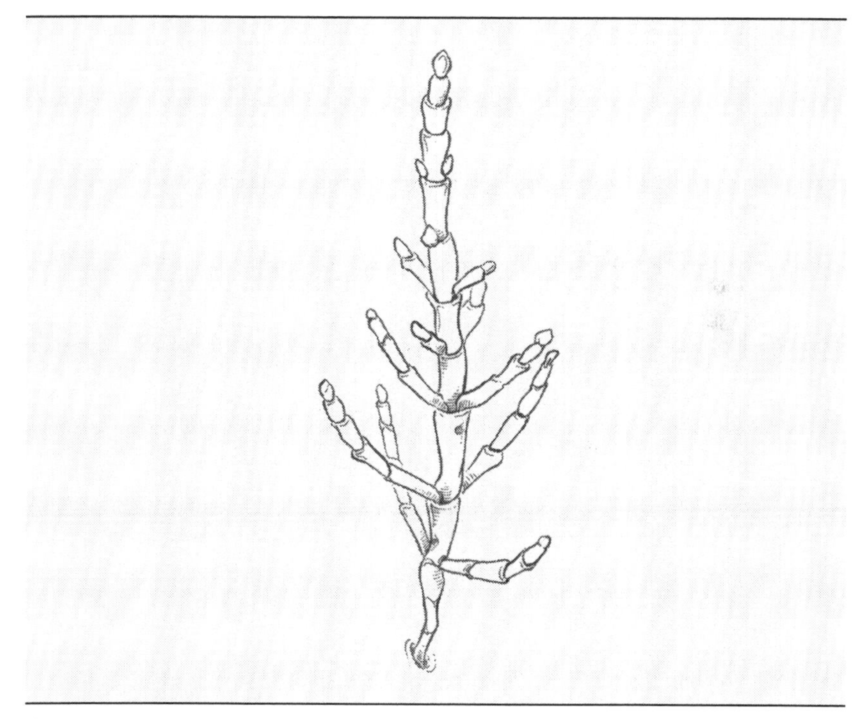

Salicornia europaea

tolerance relative to other agricultural plants, which allows it to grow in saline conditions that would be toxic to other major agricultural crops.[16] Six years of field trials undertaken in 1991 to test the limits of the plant's survivability in hostile conditions found that salicornia could thrive in extreme coastal desert conditions using seawater as its only irrigation source. Remarkably, it could produce greater yields of seeds and biomass under these conditions than soybeans grown under ideal agricultural conditions.[17] Originally conceived as an alternative to soybeans for chicken feed, salicornia's ability to thrive in marginal, non-agricultural lands makes it attractive as a non-food biomass feedstock. Furthermore, the ability to irrigate it with seawater provides an opportunity to cultivate dry coastal lands for biofuels production without depleting freshwater sources.

Salicornia faces many challenges similar to those facing jatropha as an agricultural crop. While edible, salicornia has never been domesticated in the same manner as other feedstocks such as soybeans and corn, making yields attractive yet erratic. Furthermore, at roughly 100 gallons of biodiesel per acre, salicornia's typical yield is far inferior to that of feedstocks such as rapeseed, jatropha, and palm oil (although twice that of soybeans).[18] There is little data to estimate the cost of fuels from salicornia although one study estimates salicornia-based biodiesel could be produced for roughly $100 per barrel.[19] Not enough scientific information is currently available on the plant to accurately assess its potential as a biofuel feedstock.

Microalgae, also known as microphytes, are photosynthetic single-cell microorganisms that produce large amounts of lipids when they are grown under conditions that deprive them of key nutrients.[20] There are as many as 50,000 species of microalgae, which grow over a wide range of temperatures in both saltwater and freshwater environments.[21] With up to 75% of the dry biomass consisting of lipids, microalgae are highly productive with oil yields as high as 6,000 gal/acre in field trials and 15,000 gal/acre in laboratory trials.

Microalgae's use as a renewable fuel feedstock was first explored by the U.S. government during the 1970s as part of its Aquatic Species Program (ASP) at the National Renewable Energy Laboratory (NREL).[22] The program was shut down in 1996 as low petroleum prices and commercialization of grain ethanol made microalgae an unattractive fuel option. Starting in 2006, however, the U.S. government launched a new era of research and development into algal biofuels. Private investors also took note, making algal fuels one of the fastest growing sectors of the biofuels industry in the last few years.[23, 24, 25]

Several factors have converged to generate this recent interest in algal fuels. An area roughly the size of Maryland could grow all the algae needed to meet U.S. demand for transportation fuels,[21] a significant advantage at a time of concern regarding the land-use constraints of other biofuel feedstocks. The potential yields of algae-based fuels are 1,100-15,000 gallons per acre compared to only 620-930 gallons per acre for grain ethanol. Furthermore, algae can be grown on marginal cropland or even within natural bodies of water, allowing its production to complement conventional crop production rather than replace it.

Algal fuel production can also improve air and water quality. Production of microalgae requires a concentrated source of CO_2, which can be provided from the exhaust of power plants, thus reducing greenhouse gas emissions to the atmosphere. Furthermore, some companies propose to grow microalgae in wastewater, a process that sufficiently cleans the discharged water to meet government drinkability standards.[26] A biomass feedstock whose production inherently contributes to cleaner air and water and requires less land than ethanol production seems the perfect solution to the many controversies surrounding biofuels. Algal fuel's popularity will likely only increase as companies such as Sapphire Energy – in which both Bill Gates and the Rockefeller family have invested heavily - continue to predict that they will be capable of producing algal fuel on a commercial scale for roughly the same price per barrel as crude oil in the near future.[27]

The industry faces several hurdles before commercialization. The most formidable is the cost to build the expansive system of ponds or closed reactors to combine the sunlight,

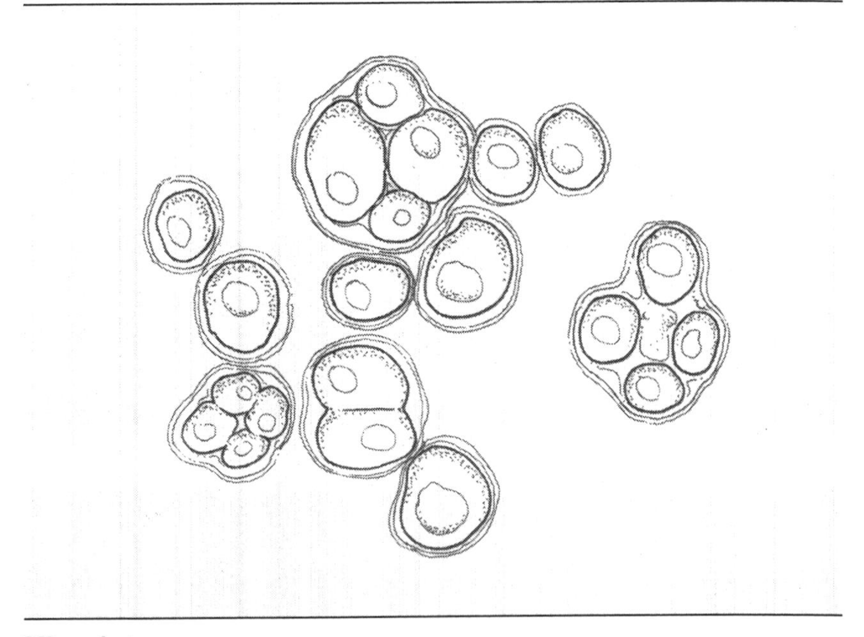

Microphytes

carbon dioxide, water and nutrients needed by microalgae to produce lipids. Open ponds, although little more than concrete raceways that continuously circulate water and screen out microalgae, are estimated to cost as much as $100,000 per acre. Photobioreactors are more elaborate closed systems suitable for highly productive pure cultures of microalgae but cost as much as $1 million per acre.

Operating costs are high because the algae must be constantly fed and harvested. The result is algal fuel that costs $20/gal or more to produce,[28, 29] although the head of the Algal Biomass Organization predicts that algal fuels will be cost competitive with petroleum by 2018.[30] This high cost is largely due to the expense of growing and harvesting microalgae, with feedstock costs being responsible for up to 60% of the cost of algal fuels.[31] One recent review finds the median projected cost of algal feedstock (in the form of triglycerides) to be $16.25/gal, with the mean higher yet.[32] This cost in particular will need to significantly decline if algal fuels are to compete with petroleum fuels in the future, even with the support of government subsidies.

Although microalgae is often touted as a solution to the "food v. fuel" debate, ironically microalgae could prove to be important source of nutrients and dietary protein for direct human consumption or indirectly through its use for animal feed. Microalgae produce large quantities of essential fatty acids (EFAs) such as Omega-3, which have been commercially exploited in human nutritional supplements and additives to poultry feed.[33] Algae are also rich in protein that might one day help meet the growing demand for protein rich foods around the world.

Neither are algal fuels a clear solution to land use concerns. Although microalgae is an aquatic species with potential to be grown in salt marshes or even open oceans, current development focuses on terrestrial systems that would compete with other land uses including irrigated food crops on arid lands that are otherwise fertile.

Although algal biofuels offers long-term prospects for efficient biomass production that would not complete with conventional agricultural land use, it remains the most expensive biomass feedstock among the leading candidates. However, as subsequently described, it is relatively straightforward to upgrade lipids, further adding to its attractiveness if algal feedstocks can be produced inexpensively.

Processing lipids into diesel fuel substitutes

Triglycerides, in principal, can be directly used as fuel in diesel engines. However, their high viscosity results in frequent fouling of the injectors and rings in diesel engines. A more reliable engine fuel results if the triglycerides are broken down into smaller molecules. Triglycerides are a class of compounds known as esters, formed by the reaction of an alcohol with fatty acids. In the case of triglycerides the alcohol is glycerol, which can bond to three fatty acids. In a process known as transesterification, the fatty acids are stripped from the backbone of glycerol and reacted with alcohols only capable of bonding with single fatty acid molecules to form three relatively small ester molecules. Usually the alcohol is methanol resulting in fatty acid methyl ester (FAME), the main constituent of biodiesel.

The chemical relationship to diesel fuel is relatively superficial. The name comes from the fact that methyl esters are plant-based compounds with combustion properties similar to fuels designed for use in diesel engines. Biodiesel can be substituted for diesel fuel in many applications although it has some prominent shortcomings. Like the lipids from triglyceride methanol glycerol methyl ester

$$\begin{array}{l} R1{-}C({=}O){-}O{-}CH_2 \\ R2{-}C({=}O){-}O{-}CH \\ R3{-}C({=}O){-}O{-}CH_2 \end{array} + 3\,CH_3OH \longrightarrow \begin{array}{l} HO{-}CH_2 \\ HO{-}CH \\ HO{-}CH_2 \end{array} + R1{-}C({=}O){-}O{-}CH_3 + R2{-}C({=}O){-}O{-}CH_3 + R3{-}C({=}O){-}O{-}CH_3$$

Transesterification of triglycerides with methanol to yield methyl esters and glycerol.

which they are obtained, fatty acid methyl esters are subject to microbial or oxidative attack, making them unsuitable in applications requiring long-term fuel storage. However, unless very tight quality control is maintained in manufacturing plants, low temperature performance of biodiesel is compromised. Another disadvantage is that the glycerol by-product of biodiesel production is contaminated with salts and other compounds, making it a low value byproduct stream.

For these reasons, transesterification is likely to be replaced by hydrogenation to convert lipids into advanced biofuels. Hydrogenation is the process of adding hydrogen to organic compounds under high pressure and in the presence of a catalyst. Hydrogenation encompasses a number of chemical reactions including breaking large molecules into smaller molecules, reducing unstable carbon-carbon double bonds into stable single bonds, rearranging molecular structures, and removing undesirable atoms such as sulfur, nitrogen, and oxygen from organic compounds. In the case of triglycerides, not only are the fatty acids removed from the glycerol backbone, but these compounds are converted into highly desirable hydrocarbon fuel molecules. Hydrogen added to glycerol forms propane, an important gaseous fuel. Hydrogen added to fatty acids has two important effects. First, the oxygen atoms responsible for the acidic nature of fatty acids are removed, transforming these long-chain molecules into hydrocarbons. Second, the carbon-carbon double bonds in these long chains, which are responsible for the tendency of lipids to go rancid over time, are converted to single bonds, producing a saturated hydrocarbon with improved stability as a fuel. The nature of these hydrocarbons depends upon the nature of the lipid feedstock. Rapeseed, for example, produces mostly hydrocarbon chains containing 16, 18, or 22 carbon molecules. Thus, these hydrocarbons can rightly be described as diesel fuel since petroleum-based diesel contains hydrocarbons containing between 10 and 22 carbon atoms (although there are some differences in the arrangement of these carbon atoms).

Many companies have demonstrated the hydrotreatment of triglycerides from vegetable oils and animal fat to produce synthetic diesel fuel including BP, ConocoPhillips, Neste Oil, Petrobas, Syntroleum, and UOP.[34] The propane and other light hydrocarbons produced are used to generate hydrogen needed by the hydrotreater. Wide scale commercial deployment awaits greater supplies of low-cost lipid feedstocks.

The use of refinery hydrotreaters presents important advantages over other approaches to advanced biofuels since it reduces the capital cost of biorefinery infrastructure and allows accelerated deployment of the technology if sufficient

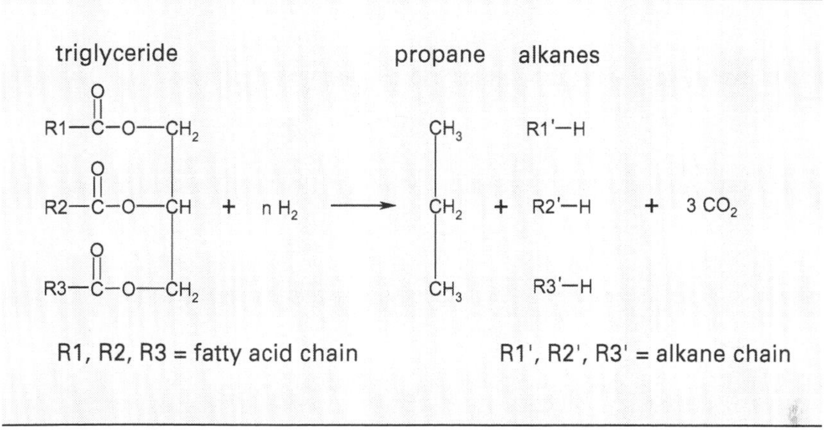

triglyceride → propane + alkanes

$$\begin{array}{l} R1{-}C(=O){-}O{-}CH_2 \\ R2{-}C(=O){-}O{-}CH \\ R3{-}C(=O){-}O{-}CH_2 \end{array} + n\,H_2 \longrightarrow \begin{array}{l} CH_3 \\ CH_2 \\ CH_3 \end{array} + \begin{array}{l} R1'{-}H \\ R2'{-}H \\ R3'{-}H \end{array} + 3\,CO_2$$

R1, R2, R3 = fatty acid chain R1', R2', R3' = alkane chain

Simplified representation of hydrogenation of triglyceride during hydrotreating.

quantities of lipid feedstock can be procured. It also raises concerns among those who imagined lipid-based biofuels would always be the province of farmer-entrepreneurs. However, even the first generation of lipid-based manufacturing plants have shifted over time from a few million gallons per year to several tens of millions of gallons capacity including a couple of plants in the U.S. of about 100 million gallons capacity. Economies of scale are as important to biorefineries as they are for petroleum refineries although the optimal sizes are likely to be quite different because of constraints in moving solid, low-density biomass to processing facilities.[35]

Other approaches for obtaining energy from algae

Like oilseeds, microalgae accumulate lipids within the cells where it represents an energy store for the organism. But unlike oilseeds which are collected once at the end of a growing season and simply pressed to release the oil, biomass grown in aqueous environments present special challenges to lipid recovery. Microalgae especially thrive under low population densities. High yields of oil are made possible by continuously harvesting low density suspensions of microalgae from their aqueous environment. Dewatering these low density suspensions requires a 500-fold concentration of the microbial biomass, which represents one of the major challenges in production of algae-based fuels.

An alternative to directly harvesting the algae is to anaerobically digest them into a more easily recoverable fuel. Anaerobic digestion is the decomposition of organic waste into gaseous products by bacteria in an oxygen-free, aqueous environment.[36] It is widely used to treat municipal solid waste, municipal and industrial waste water and

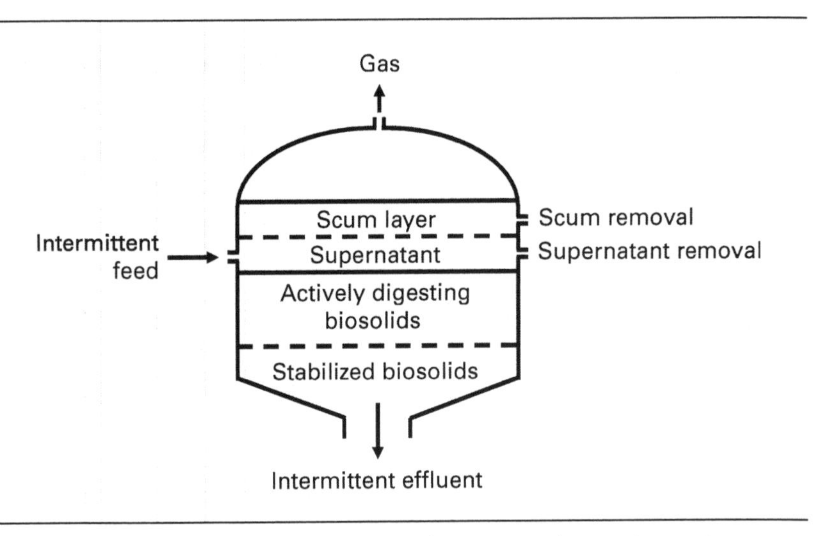

Biogas from Microalgae. An anaerobic digester continuously produces mixture of CH4 and CO2 from an intermittent charge of algae suspended in water. *Source: Reference 41.*

increasingly animal manure for the purposes of pollution control. The process successively breaks down organic matter into simpler compounds. The desired product, known as biogas, is a mixture of CH_4, CO_2 and some trace gases. Most digestion systems produce biogas that is between 55% and 75% methane by volume. Typical efficiency in converting the chemical energy of dry matter to methane is about 60% with methane yields ranging between 8,000-9,000 cubic feet/ton of volatile solids added to the digester.[37] However, thermodynamic efficiencies approaching 90% with methane yields of 11,000–13,000 cubic feet/ton volatile solids have been achieved in research programs.

Biogas, once treated to remove sulfur compounds, can

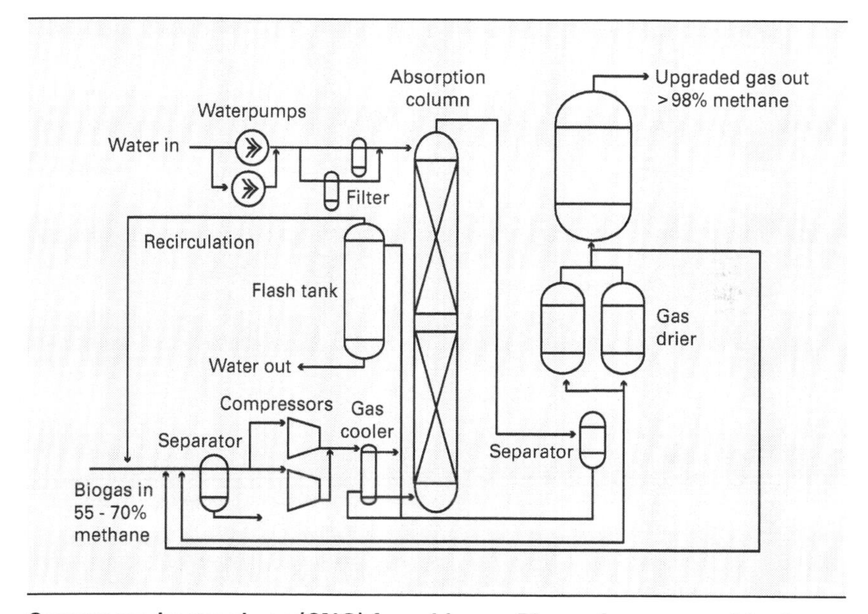

Compressed natural gas (CNG) from biogas. Biogas from anaerobic digestion of microalgae is upgraded to CNG by compression, removal of CO2 and trace gas contaminants, and drying. *Source: IEA Bioenergy, Task 37.*

substitute for natural gas in many applications, including stationary power generation. Biogas can be further purified into synthetic natural gas (SNG) by removing the CO_2. Some communities have experimented with converting the biogas into compressed (synthetic) natural gas (CNG) to serve as transportation fuel.[38]

The expediency of anaerobically digesting microalgae biomass was described as early as 1957 by Golueke and colleagues at the University of California-Berkeley.[39] Anaerobic digestion of algae was found to be slower and less complete than for raw sewage sludge yielding less gas product per pound of volatile matter. The result was somewhat surprising as the high fatty acid content of the algae should have made it an ideal growth substrate. Golueke's team hypothesized that either the high ammonia content of the substrate or the resistance of live algae cells to bacterial attack were responsible. In an effort to kill the algae, they reran digestion trials at temperatures as high as 50°C, which substantially improved the digestion of algae. In the 1980's, Samson and LeDuy recognized that the high ammonia content of microbial biomass meant that the carbon to nitrogen ratio was not in proper balance for optimal biological activity.[40] They added carbon-rich wastes to the media and observed significant improvement in both yield and productivity of methane. Anaerobic digestion of the microalgae also offers a solution to the problem of supplying microalgae with sufficient quantities of CO_2 for their growth. In 1959 Golueke and his collaborators[41] proposed and tested a "closed" system for converting solar energy into methane fuel. It consisted of a photosynthetic chamber that provided water, CO_2, ammonia, and other nutrients for the growth of microalgae. The microalgae, in turn were anaerobically digested in a separate chamber to produce methane along with the co-products CO_2, ammonia, and other nutrients. These co-products were returned to the photosynthetic chamber to support growth of additional microalgae. The overall energy efficiency of converting sunlight into fuel was about 2%, which is comparable to cellulosic-based bioenergy systems when the efficiency of photosynthesis is included in the analysis.

9

How can we turn cellulose into advanced biofuels?

The nature of cellulosic biomass

Cellulose, although less energy dense than lipids, is the most abundant biomass on the planet. This plant polymer dominates most natural ecosystems and it has been cultivated for millennia as fodder for livestock. As the name suggests, molecules of these 'hydrates of carbon' often contain equal molar quantities of carbon and water (CH_2O).

The most familiar carbohydrates are simple sugars like glucose and fructose, known as monosaccharides, and sucrose (table sugar), a disaccharide formed from glucose and fructose. Many monosaccharides and disaccharides can be fermented to ethanol, which is the basis of the sugar cane ethanol industry. However, the most plentiful carbohydrates in nature are long polymeric chains of simple sugars such as starch, hemicellulose, and cellulose, which in general cannot be directly fermented. These polysaccharides must first be broken down by reaction with water in the presence of acids or enzymes to form monosaccharides. This hydrolysis of carbohydrate has been mastered for starch and is the basis of the grain ethanol industry.[1]

Most cellulose occurs in nature as a composite material called lignocellulose or simply plant fiber.[2] This composite consists of cellulose fibers surrounded by another sugar polymer called hemicellulose and imbedded in a matrix of phenolic polymer called lignin. Separating these three polymers so that the carbohydrate content can be hydrolyzed to simple sugars is one of the major challenges of advanced biofuels research.

Lignocellulosic biomass of interest as feedstock for biofuels production is generally classified as either wastes or dedicated energy crops.[3] A waste is a material that has been traditionally discarded because it has no apparent economic value or represents a nuisance or even a pollutant to the local environment. Categories of waste materials that qualify as cellulosic biomass include municipal solid waste, agricultural processing byproducts, and sometimes even agricultural residues. Municipal solid waste (MSW) is whatever is thrown out in the garbage and clearly includes materials that do not qualify as biomass, such as glass, metal, and plastics. Food processing waste is the effluent from a wide variety of industries ranging from breakfast cereal manufacturers to alcohol breweries. Obviously, they vary greatly in their cellulosic content as well as other properties important to their conversion to fuel such as moisture content and energy content. Agricultural residues are simply that part of a crop discarded after harvest. Corn stover, which is the stalks, leaves, husks, and cobs of the corn plant, is the largest agricultural residue in the United States.

Despite the size of this resource, corn stover has traditionally been left in the fields. Conventional tillage plows stover into the ground ahead of the next planting season while no-till agriculture leaves it on the surface to decay. It is sometimes collected from the field and baled after the grain has been harvested to use as animal fodder. Left on the fields, it plays a role in controlling soil erosion, building soil organic matter, and recycling nutrients to the soil although it is generally acknowledged that some portion of the stover can be harvested, the amount depending upon several factors including climate, topography, soil type, and tillage practice.[4]

Corn stover has several attractive features as biofuels feedstock. First, as a byproduct of conventional agriculture, its use as a biofuel feedstock does not divert crops away from food production. Nor does its production require conversion of grassland or forest to agriculture. A corn crop yields about one ton of stover for every ton of grain.[5] With the U.S. corn crop averaging 11.5 billion bushels in recent years,[6] this translates into almost 300 million dry tons of stover. Recent studies suggest that almost 120 million dry tons could be annually collected without compromising soil quality.[4] As subsequently described, advanced biomass conversion techniques are expected to produce between 75

Corn stover

– 100 gallons of ethanol from cellulosic biomass. Thus, corn stover could provide around 10 billion gallons of ethanol per year – about the same amount as currently produced from corn grain in the United States.

Dedicated energy crops are cellulosic biomass grown specifically for production of fuels, chemicals, and power; that is, for purposes other than food or feed. Harvesting may occur on an annual basis, as with switchgrass, or on a 5-7 year cycle, as with certain species of fast-growing trees such as hybrid poplar. Dedicated energy crops are conveniently divided into herbaceous energy crops (HEC), like switchgrass or miscanthus, and short rotation woody crops (SRWC),[7] like hybrid poplar or willow. Although ethanol yields (gallons per ton) from energy crops are expected to be lower than from corn grain, biomass productivities (tons per acre) are expected to be higher resulting in much higher ethanol productivity per acre of land.

Another advantage of growing dedicated energy crops compared to conventional agricultural crops such as corn,

Energy crops have lower environmental impacts compared to conventional crops.

	Reduction in agricultural intensity metrics for energy crops relative to average inputs for corn-wheat-soybean production	
Metric	Woody biomass	Herbaceous biomass
Erosion	12.5-fold	125-fold
Fertilizer	2.1-fold	1.1-fold
Herbicide	4.4-fold	6.8-fold
Insecticide	19-fold	9.4-fold
Fungicide	39-fold	3.9-fold

Source: Reference 8.

wheat, and soybeans is better environmental performance.[8] Erosion rates are expected to be ten to one hundred times lower for energy crops because most do not require annual tillage of the ground, which exposes the soil to wind and water-driven erosion. Perennial plants are also more conservative of nutrients, reducing fertilizer demand by as much as one-half and, once established, are able to out-compete weeds, reducing herbicide use by a factor of four to seven. Insecticide and fungicide applications are often reduced by a factor of ten or more.

Finally, energy crops have vast potential to be improved through plant breeding. Conventional crops have been bred for generations to increase sugar, starch, protein, and lipids at the expense of cellulose. For example, the Green Revolution of the twentieth century intentionally sought to shorten the length of rice stalks to reduce lodging and channel more photosynthetic energy into grain production.[9] Shifting the focus to breeding plants with high levels of structural polysaccharides (cellulose and hemicellulose) instead of storage polysaccharides (starches) could dramatically increase biomass yields. Molecular biology-based techniques introduced in the last few years, including transgenic manipulation of plants, can be expected to accelerate plant breeding for this purpose.[10]

Herbaceous crops are plants that have little or no woody tissue. Herbaceous crops include both annuals and perennials. Annuals die at the end of a growing season and must be replanted in the spring. Perennials die back each year in temperate climates but reestablish themselves each spring from rootstock. Perennials are generally preferred for energy crops because, once established, they require less labor and agricultural inputs such as fertilizer and pesticides. Herbaceous plant material contains less lignin than woody plant material, which increases the yield of ethanol per ton of biomass. Lower lignin content also eases the extraction of cellulose and its conversion to sugars. Two of the most prominent herbaceous plants being considered for cultivation as dedicated energy crops are switchgrass and Miscanthus.

Switchgrass (*Panicum virgatum*) is a native prairie grass, one of the plant species that dominated the prairies of the

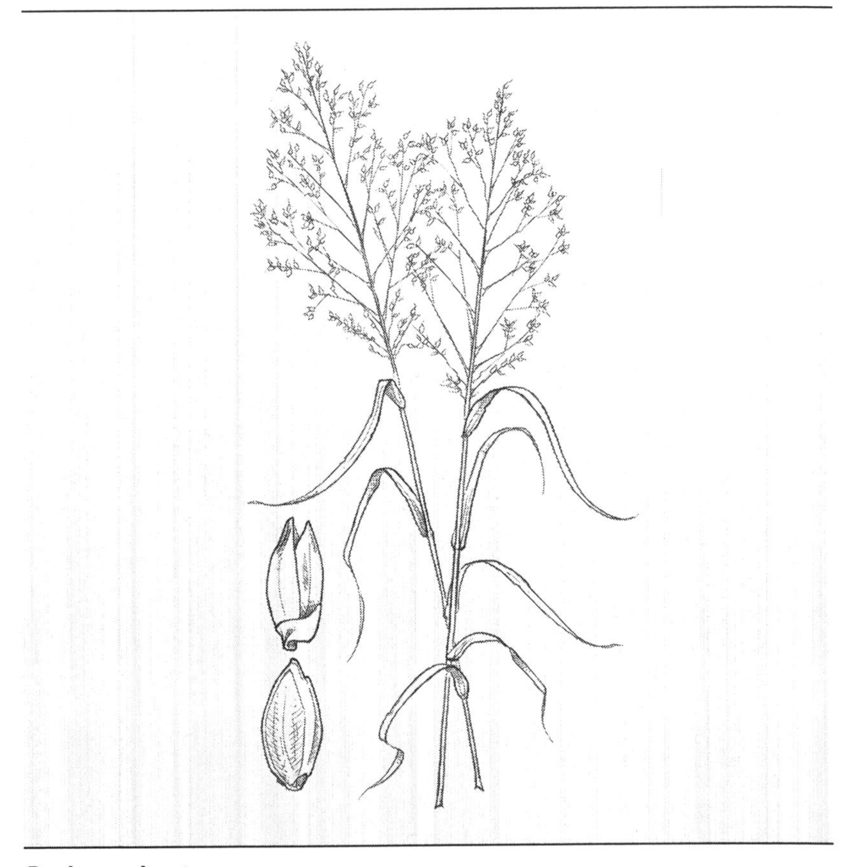

Panicum virgatum

North American continent until the 19th century when these natural landscapes were converted to agriculture.[11] Switchgrass, like the other native prairie grasses, was a nuisance for farmers, as its significant height (2.5m), heavy sods, and ability to establish itself from both seed and root stock made it difficult to clear. Those very same attributes make it an ideal cellulosic ethanol feedstock today. Its above ground biomass can exceed 5 tons per acre,[12] while its deep roots allow it to thrive with minimal fertilizer application. Finally, its reseeding ability and status as a perennial grass mean that, unlike crops such as corn and soybeans, it can be harvested annually over a period of several years without replanting. The suitability of switchgrass as a biofuel feedstock was recognized by researchers in the mid-1980s and, as a result, an abundance of scientific data is available. Current technology allows for biofuel yields of 70 gallons per ton[13] which, while less than that of corn grain per ton of feedstock, combines with switchgrass' greater biomass production per acre to yield more ethanol per acre than corn.[14]

Giant miscanthus (*Miscanthus x giganteus)* is a warm-season Asian perennial grass capable of growing to 2-3 meters in height. It is sterile and only propagated through planting of root stock (rhizome propagation), preventing it from causing a non-native invasion[15] - an important trait for any potential non-agricultural feedstock. It can yield up to 16 tons per acre with the use of little to no fertilizer and pesticides.[5] Unlike some other perennial grasses, Miscanthus can be harvested in the winter or early spring because the thick stalks resist falling over (lodging) during wind storms or snow falls.

Miscanthus has not yet been cultivated on a large scale, making ultimate yields difficult to estimate. Thus far smaller field tests have generated the largest yields, with those amounts decreasing as the plot size increases. The capital expenditures associated with establishing Miscanthus fields are also substantial at this time, although these are expected to drop as commercial establishment becomes more prevalent and more efficient techniques are developed.

Short rotation woody crops (SRWC) are fast-growing woody biomass that is suitable for use in dedicated feedstock supply systems.[7] Desirable SRWC candidates display

Miscanthus x giganteus

rapid juvenile growth, wide site adaptability, and pest and disease resistance. Woody crops grown on a sustainable basis are harvested on a rotation of 3 - 10 years. Woody biomass contains more lignin than herbaceous biomass, which is both an advantage and disadvantage. Lignin contributes to the strength and pest resistance of plant cell walls, which allows woody biomass to be stored more reliably than herbaceous biomass. However, higher lignin content results in less carbohydrate in the biomass. Woody crops include hardwoods and softwoods. Hardwoods are trees classified as angiosperms, which are also known as flowering plants. Hardwoods can regrow from stumps, a process known as coppicing, which reduces their production costs compared to softwoods. Softwoods are trees classified as gymnosperms, which encompass most trees known as evergreens. Softwoods are generally fast growing but their carbohydrate is not as accessible for chemical processing as the carbohydrates in hardwood.

One of the most prominent woody crops being considered for cultivation as a dedicated energy crop is hybrid poplar. A cross between American cottonwood and European poplar trees, the hybrid poplar (*Populus deltoides x Populus nigra*) is a fast-growing tree that can attain full height in 7 to 8 years. Its rapid growth and ability to thrive under arid conditions makes it a potential feedstock for cellulosic biofuels. Its feedstock potential was first realized during the Oil Shocks of the 1970s when a consortium of researchers from government agencies, universities, and the private sector began studying it. One result of this research was the complete sequencing of the tree's genome, allowing for the optimization of its feedstock capabilities.[16] Like many woody crops, hybrid poplar faces increased timescales to recover investment costs. Whereas traditional biofuel crops are harvested months after planting, hybrid poplar must be allowed to grow for a few years at minimum before harvesting if optimal yields are to be attained.

In 2005 the U.S. Department of Agriculture sponsored a study on potential biomass supply in the U.S., including both dedicated crops and agricultural residues.[17] An update to this study was published in 2011.[18] The conclusion of this study and its update is that in excess of 1 billion tons of dry biomass, representing 25 billion GJ of energy, could be produced in a sustainable manner by 2030.

The land base of the U.S. encompasses nearly 2,263 million acres. About 33% is classified as forestland; 26% as grassland, pasture and range; 20% as cropland; 8% special

Populus deltoides x Populus nigra

uses (e.g., public facilities); and 13% miscellaneous such as urban areas, swamps, and deserts.[19, 20] The USDA study considered how the first three categories of land could be employed in biomass production, carefully excluding inaccessible and environmentally sensitive lands from consideration. It also accounted for use of these lands in the production of conventional forest products and agricultural commodities.

The study included some forward looking projections on agricultural technology: increases of small grain yields by 50%; residue ratio for soybeans increases to 2:1; harvest technology recovers 75% of crop residues (when sustainable removal is possible); all cropland is managed by no-till methods; 55 million acres of cropland, idle cropland, and

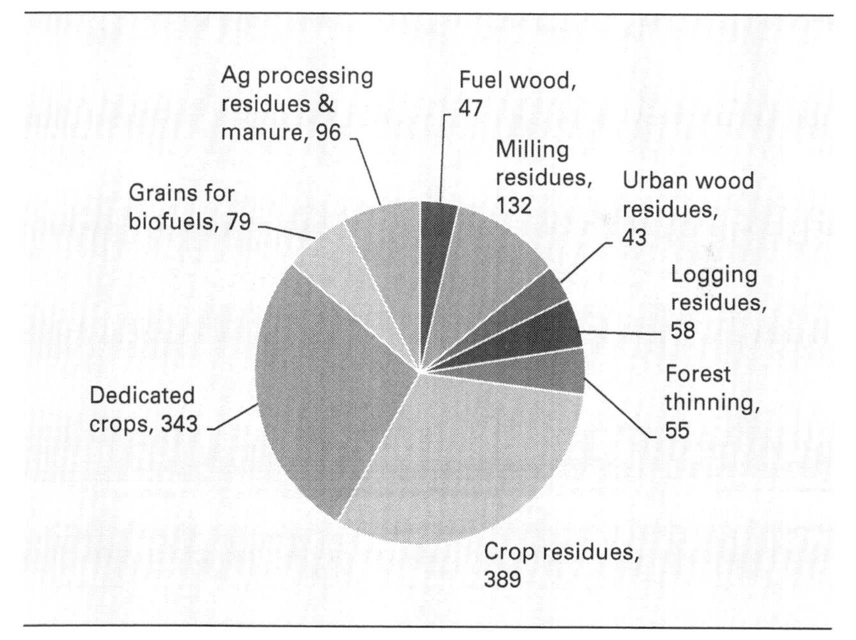

U.S. Biomass Potential is 1.2 billion tons (values in millions of metric tons).
Source: Reference 17.

pasture are dedicated to perennial bioenergy crops; and all manure in excess of allowable on-farm soil application is used for biofuel.

Forestry resources, in the form of fuel wood, milling residues, urban wood residue, logging residues, and wood recovered from forest thinning (for forest fire control) could yield 335 million Mg of dry wood per year. Agricultural resources, in the form of crop residues, perennial crops, grains for biofuels, and processing residues and manure, could yield 907 million Mg of dry biomass per year. Crop residues and perennial crops represent 31% and 28%, respectively, of the total biomass supply. Assuming an average heating value of 18 MJ/kg, this 1.2 billion Mg supply of biomass represents 21 billion GJ of energy, or 21% of total U.S. energy consumption. The report notes that this is sufficient supply to meet the U.S. Department of Energy's goal of producing one-third of U.S. transportation fuel from biomass. This conclusion has commonly been misinterpreted as a ceiling on biofuels production. In fact, biofuels researchers hope to eventually produce 100 gallons or more of biofuel per ton of biomass. Even if this biofuel was ethanol, which only has two-thirds the heating value of gasoline, 1.3 billion Mg of biomass could yield over 85 billion gallons of gasoline-equivalent biofuel, which is almost 60% of current gasoline consumption. Thus, biomass could represent a significant contribution to the future U.S. energy supply.

Getting sugars out of biomass

Cellulose is a structural polysaccharide consisting of a long chain of glucose molecules linked by glycosidic bonds.[21] Breaking the links in the chain releases the glucose and makes it available for either food or fuel production. Although very similar to starch in chemical composition, cellulose is digestible by relatively few kinds of animals. Cattle and termites have figured out how to breakdown cellulose through symbiotic relationships with bacteria in their guts, allowing them to consume cellulose-rich grass and wood, respectively. The bacteria produce enzymes called cellulases, which snip the glycosidic bonds between the glucose molecules in a process known as enzymatic hydrolysis.

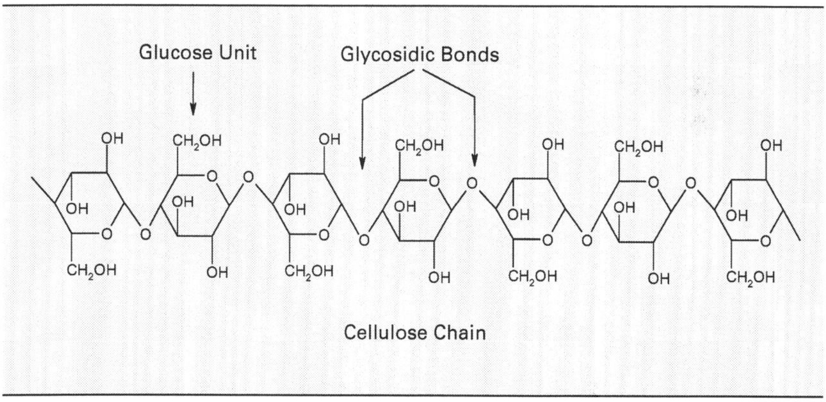

Cellulose is a long chain of glucose units connected by glycosidic bonds.
Source: Author.

Lignocellulose is designed by nature as structural material for plants that is resistant to microbial attack. Before cellulose in grass or wood can be hydrolyzed to glucose, it must be partially released from the lignin matrix to make it accessible to cellulases secreted by bacteria. The placid "chewing the cud" by cattle is a form of biomass preprocessing to facilitate cellulosic hydrolysis and release of glucose and other simple sugars from the polysaccharides in biomass. Despite these evolutionary advances, only 50-80% of the energy content of cellulosic biomass is captured by cattle, depending upon the lignin in the forage, because some of the polysaccharides and all of the lignin pass through the gut undigested.[22]

Biochemical processing uses enzymes to release simple sugars from biomass in a manner that replicates the process in the guts of cattle and termites.[23] Pretreatment breaks up plant fibers, making the polysaccharides more susceptible to hydrolysis. Enzymatic hydrolysis breaks the polysaccharides into simple sugars. Fermentation biologically converts these

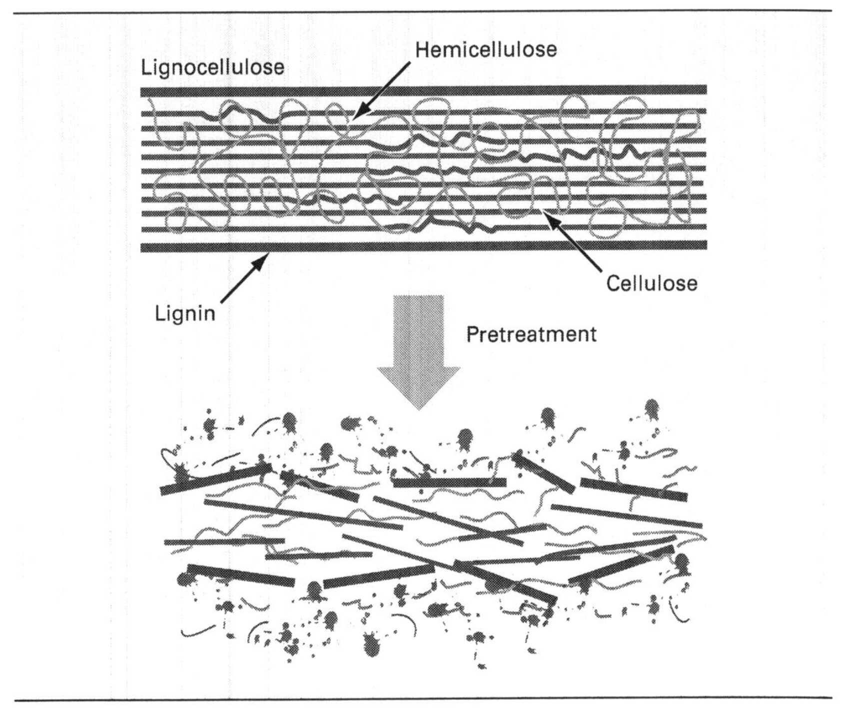

Lignocellulose is a natural composite of cellulose, hemicellulose, and lignin. It resists biological degradation unless pretreatment is able to make the cellulose more accessible to cellulose-degrading enzymes.
Source: Author.

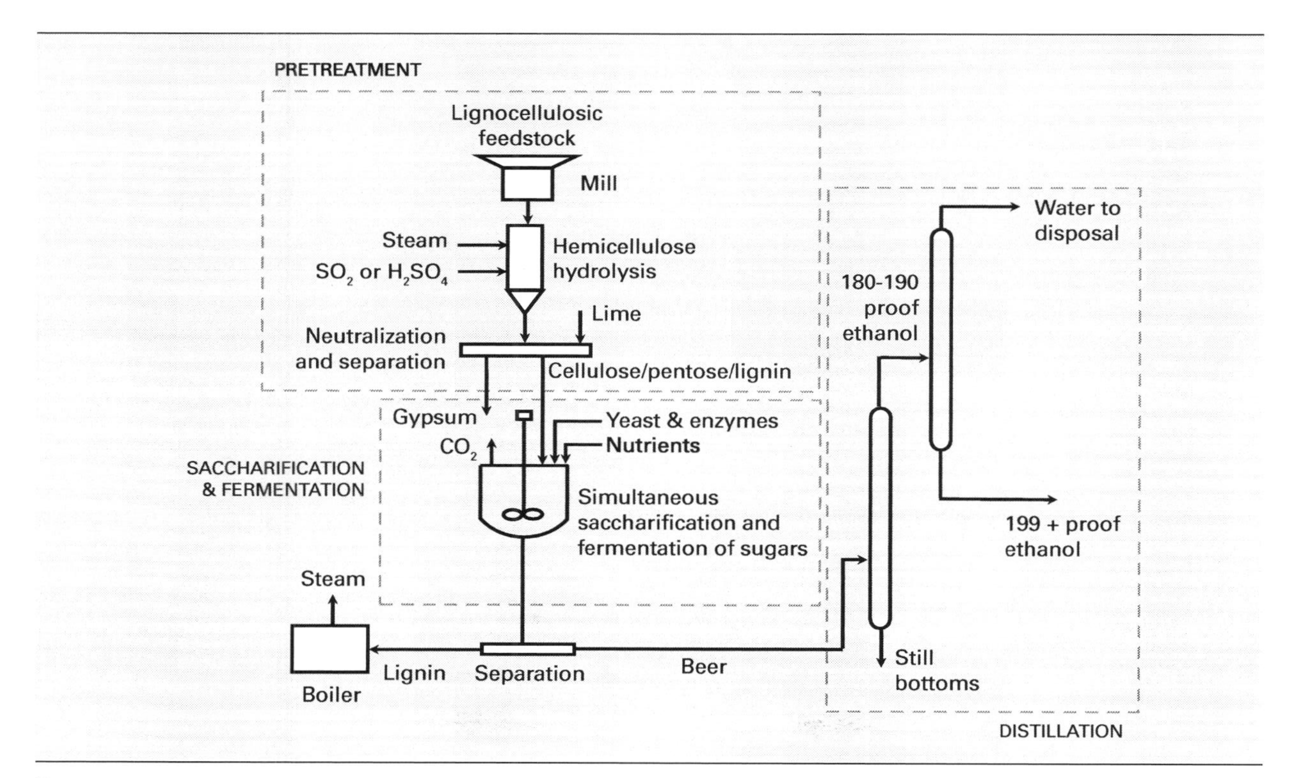

Four steps to cellulosic ethanol: Pretreatment, saccharification, fermentation, and distillation.

Source: Adapted from Reference 23.

simple sugars into ethanol. Distillation separates the ethanol from water and unconverted biomass.

Pretreatment begins by grinding up the biomass although this is not enough to release cellulose and hemicellulose from the polymeric matrix of lignin. Various thermal and chemical processes have been devised to disrupt the lignocellulose structure by poorly understood mechanisms. Among the most common is treatment with steam and sulfur dioxide or sulfuric acid, which breaks down the hemicellulose to five carbon sugars (pentose) and smaller quantities of six carbon sugars (hexose) and in the process, opens up the structure of cellulose making the fibers more susceptible to subsequent processing steps. The acidic mixture of cellulose fibers, sugars, and lignin is treated with lime to neutralize the solution ahead of saccharification and fermentation. Pretreatment is the most costly step in cellulosic ethanol production, accounting for about 33% of the total processing costs.

Saccharification is the release of simple sugars from polysaccharides by physical, chemical, or biochemical processes.[24] Often the pretreatment process accomplishes this for hemicellulose, releasing a variety of simple sugars. As cellulose is more recalcitrant, much of the focus of saccharification research has been on breaking down cellulose to the simple sugar glucose. Cellulose saccharification is usually accomplished through hydrolysis, the reaction of water to break glycosidic bonds between glucose units in the polysaccharide chain. Although acid hydrolysis was developed early in the twentieth century, much of the current focus is on enzymatic hydrolysis, which has scope for significant improvements through advances in biotechnology.

Enzymatic hydrolysis of cellulose proceeds in several steps by the action of a system of enzymes known as cellulases.[25] The process begins with the liberation of cellulose

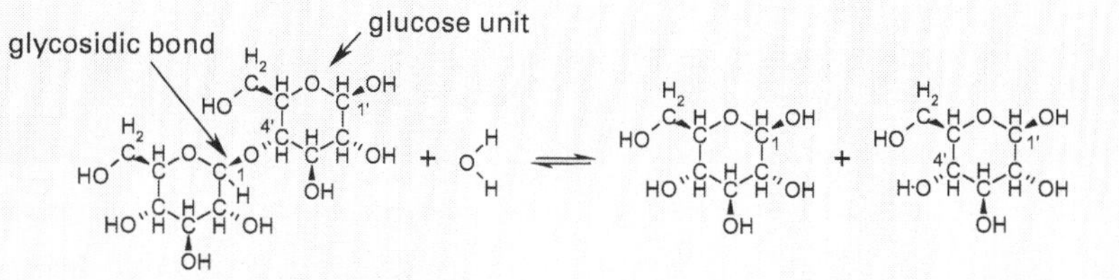

glycosidic bond
glucose unit

cellobiose (two linked glucose units) glucose monomer glucose monomer

Just add water and enzymes. Saccharification is the process of releasing simple sugars from complex carbohydrate. Enzymatic hydrolysis of cellulose breaks the glycosidic bonds between glucose molecules by the action of water in the presence of special enzymes called cellulases.
Source: Reference 23.

strand from crystalline cellulose by endocellulases that hydrolyzes glycosidic bonds internal to the cellulose structure. This is followed by cleavage of shorter chains of glucose (oligosaccharides) from the ends of the released cellulose strands. The low specific activity leads to high enzyme loading requirements: approximately 1 kg of enzyme is needed for hydrolysis of 50 kg of cellulose fibers. Conversion rates are as low as 20% in 24 hours; thus, up to seven days are required to digest lignocellulose.

Although pretreatments can hydrolyze hemicelluloses, they can also degrade the sugars released. For this reason, the use of enzymes for both cellulose and hemicellulose are under development.[26] Unlike cellulose, which consists of only one kind of monomeric sugar, hemicellulose consists of many kinds of hexose and pentose sugars. Each glycosidic bond between pairs of monosaccharides in the chain of a polysaccharide requires a hydrolysis enzyme unique to that pair. Thus, "cocktails" of hemicellulases formulated for specific types of biomass feedstocks must be formulated to release pentoses and hexoses from hemicelluloses.

Cellulases and hemicellulases are produced from a variety of fungi and bacteria although commercial production comes predominately from the fungi of the genera *Trichoderma* and *Aspergillus*.[27] Enzymes are relatively expensive to produce, once costing as much as $5.00 per gallon of ethanol although recent development efforts have reduced the price to $0.20 per gallon.[28] Nevertheless, the cost of enzymes must be dropped another factor of ten before cellulosic ethanol becomes economically feasible. For commercial

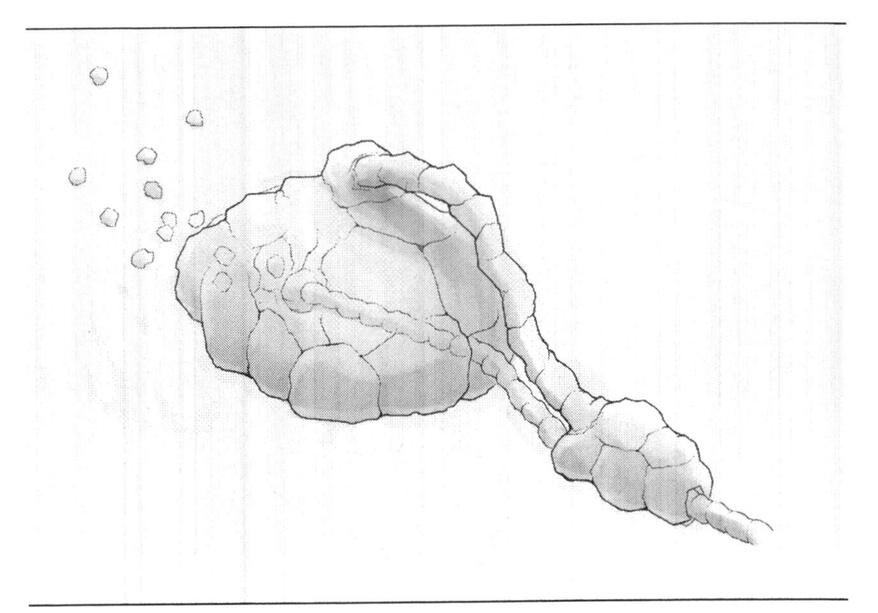

Cellulases are enzymes that decompose cellulose into simple sugars. These large protein structures detach single strands of cellulose from bundles of cellulose fiber and break the bonds between glucose units in the chain. *Source: Redrawn from illustration prepared by the U.S. DOE National Renewable Energy Laboratory.*

applications, enzymes will have to be grown on less expensive substrates and have higher volumetric productivity, higher specific activity on solid substrates, increased stability at elevated temperatures and at certain pH, and better tolerance to end product inhibition.[29]

These sugars can then be fermented to ethanol in a manner similar to the process employed in first-generation ethanol plants. The first step in a successful fermentation is removal of toxic compounds from the substrate that would otherwise inhibit the growth of fermentation organisms. These toxic compounds include furfural and acetic acid, which are breakdown products from hydrolysis of hemicellulose. Traditional detoxification methods, such as the addition of activated carbon and extraction with organic solvents can be costly. Another method under development is adaptation of the fermentation organisms to the inhibitory substances.

Numerous yeast species, including common baker's yeast, *Saccharomyces cerevisiae*, and two species of bacteria in the genus Zymomonas, efficiently ferment six-carbon sugars to ethanol and carbon dioxide.[30] However, they are not able to ferment the pentoses released from lignocellulosic biomass upon hydrolysis of the hemicellulosic fraction. Efficient conversion of lignocellulosic biomass requires fermentation of pentoses, especially xylose and arabinose. A variety of microorganisms can directly ferment pentose but yields are only on the order of 50 grams of ethanol per liter of solution compared to 150 grams per liter for common baker's yeast fermenting glucose or sucrose.[31]

Research on the use of wild-type bacteria and fungi to convert xylose waned during the 1980's, due in part to the widely recognized disadvantages of low conversion rate and/or poor yield of these types of microorganisms. Instead, recombinant DNA techniques are being employed to produce new strains of microorganisms with the desired trait of fermenting both hexoses and pentoses. Using recombinant techniques, researchers at the University of Florida transferred genes from *Z. mobilis* to *E. coli*,[32] which resulted in a microorganism able to ferment up to 90% of the sugars derived from lignocellulose with final ethanol concentrations ranging from 40–58 grams per liter. Similarly, researchers at Purdue University introduced genes for xylose utilization into *Saccharomyces cerevisiae*.[33,34]

Originally, saccharification and fermentation were carried out in separate reactors. However, the glucose released during saccharification inhibited cellulase activity, requiring lower solids loading. This had the disadvantage of reducing ethanol concentrations and increasing processing costs. Simultaneous saccharification and fermentation (SSF) was introduced as a way to overcome end product inhibition by fermenting glucose as it is formed. The optimum temperature for the SSF reactor is a compromise between the 20–32°C range required for fermentation and the 50°C temperature required for maximum cellulase activity. Pentose released during biomass pretreatment was separately fermented.

Simultaneous saccharification and co-fermentation of hexose and pentose (SSCF) is similar to the SSF process but eliminates the separate pentose fermenter, with attendant savings in capital and operating costs. The SSCF process is only possible if microorganisms are available to ferment both hexose and pentose. No microorganism is currently available for commercial realization of this process. Thus, researchers in metabolic engineering are attempting to enhance the productivity and yield of bacteria that are able to ferment both types of sugar.

Consolidated bioprocessing integrates cellulase production, cellulose hydrolysis, and pentose and hexose fermentation into a single step.[35] Advantages include the consolidation of four processing steps into a single reactor, operations are simplified, and cost of chemicals is reduced. An example of commercially successful application of CBP is anaerobic digestion of sewage sludge and agricultural wastes into methane and carbon dioxide. However, for ethanol production, yields are low, undesired metabolic by-products are produced, and product inhibition is common. Further development should improve the attractiveness of this approach.

After fermentation, the process of recovering the final product is similar for the different kinds of cellulosic ethanol production. Lignin is separated from the mixture

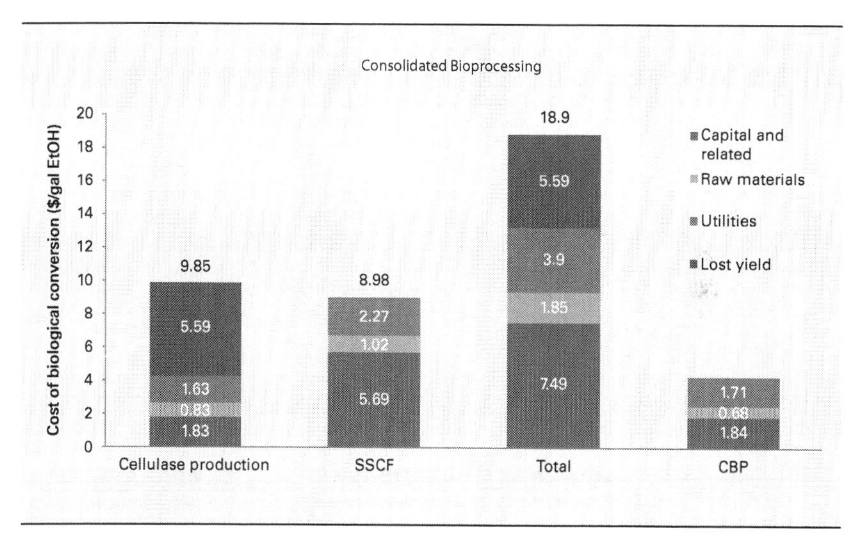

Consolidated bioprocessing (CBP) is expected to dramatically drop the cost of cellulosic ethanol. Whereas simultaneous saccharification and co-fermentation (SSCF) requires separate cellulase production, CBP combines all four processes of cellulosic ethanol production in a single reactor. *Source: Adapted from Reference 35.*

upon completion of fermentation and used as boiler fuel to produce process heat for the plant. The beer is distilled to ethanol in a process identical to that employed after sugar or starch fermentations.

The subject of intensive research for over thirty years, biochemical conversion of biomass to ethanol has not yet achieved commercial status. The challenges are many including improving pretreatments to increase the digestibility of biomass; reducing the cost of enzymes required to hydrolyze polysaccharides to simple sugars; developing microorganisms that can ferment the several kinds of sugars released from cellulosic biomass; and effectively harnessing the energy trapped in indigestible lignin.[38] In 2007 the United States embarked on a five-year, $400 million program of fundamental research at three new research centers to overcome the challenges of biochemically generating

Ethanol yields depend upon the kind of biomass used as feedstock

Feedstock	Cellulosic ethanol yields (gallons per dry ton of feedstock)
Sugar and starch crops	
Corn grain	110
Sugar cane	20
Herbaceous crops (low lignin)	
Corn stover	113
Bagasse	111
Woody crops (high lignin)	
Forest thinnings	82
Hardwood sawdust	101

Source: Sugar and starch crop data are actual ethanol yields obtained from Reference 36. Other crop data are theoretical ethanol yields obtained from Reference 37 (yields to date are as much as 40% lower).

ethanol from cellulosic biomass. That same year BP invested $500 million in an Energy Biosciences Institute led by the University of California–Berkeley to investigate biological approaches to advanced biofuels.

The limitations of ethanol as a transportation fuel has encouraged the search for alternative fermentations that can produce hydrophobic molecules that are less oxidized than ethanol. These alternatives include higher alcohols (most prominently butanol), fatty acids, fatty alcohols, esters, alkanes, alkenes, and isoprenes.

Butanol, a four-carbon alcohol with superior fuel properties compared to either methanol or ethanol, has received considerable attention as an advanced biofuel.[39] It can be fermented from sugar using the bacterium *Clostridium acetobutylicum* in a process known as acetone, butanol, and ethanol (ABE) fermentation.[40] This commercial process was eventually superseded by less expensive petroleum-derived butanol, but interest in biofuels has renewed interest in biobased butanol. Despite its advantages as fuel compared to ethanol, its production is currently more expensive than ethanol. Current butanol yields are very low, rarely exceeding 20 g/L compared to 150 g/L for traditional ethanol fermentations. Butanol is highly toxic to microorganisms, a major challenge to improving yields even through metabolic engineering. Butanol distillation is also more energy intensive than ethanol distillation.

Hydrocarbons can also be synthesized in microbial fermentations.[41] The common yeast *Saccharomyces* can synthesize a variety of straight-chain and branched-chain hydrocarbons containing between ten and thirty-four carbon atoms, which is a range suitable for upgrading to transportation fuels. Yields are a respectable 10.2 wt-% of dry biomass under anaerobic conditions.[42] The highest reported yield of hydrocarbons in wild-type bacteria is the salt-tolerant bacterium *Vibrio furnissii*, which produces hydrocarbons at 60 wt-% of dry biomass.[43] More commonly, hydrocarbon yields are just a few percent weight of dry biomass. Metabolic engineering could improve yields of hydrocarbons by mapping metabolic fluxes associated with their production and identifying enzymes responsible for hydrocarbon synthesis.

For example, terpene synthesis has been promoted in both eukaryotes and prokaryotes through metabolic engineering.[44] Unlike most biomolecules, which incorporate copious quantities of oxygen into their molecular structures, terpenes contain little or no oxygen. Terpenes are constructed from a molecular building block known as isoprene C_5H_8, which is one of a class of compounds characterized by the presence of two pairs of double carbon bonds separated by a single carbon bond (isoprenes). Both terminal carbon atoms can bond with terminal carbon atoms of other isoprenes to form a variety of hydrocarbon compounds with different degrees of saturation (also of interest are closely related terpenoids, which are derived from terpenes by moving or removing methyl groups or adding oxygen atoms). Although many plants and microorganisms have been found to produce terpenes, they are usually produced in tiny amounts, too small for economical exploitation.

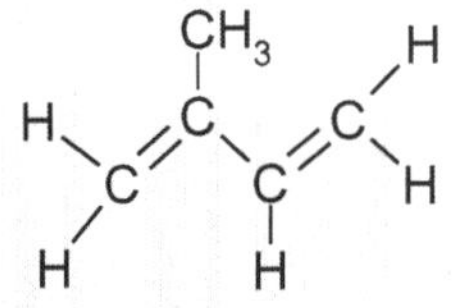

Isoprene – the building block of terpenes. *Source: Reference 45.*

As a first step in overcoming this limitation, Joseph Chappell at the University of Kentucky and Joseph Noel of the Salk Institute for Biological Studies in San Diego, California cloned the first plant enzyme responsible for terpene synthesis and elucidated its structure.[44] Subsequently they founded the company Allylix to commercially exploit these discoveries. Other researchers are using synthetic biology to assemble microorganisms specifically designed to manufacture terpenes. Amyris Biotechnologies, Inc. and LS9 are examples of companies that are commercially developing synthetic biology for advanced biofuels and chemicals manufacture. Like other forms of lipids, terpenes must be upgraded to a final fuel product. Although very little oxygen would have to be removed, saturation and structural rearrangement of the terpene molecules is required to obtain suitable gasoline, diesel fuel, or aviation fuel products.

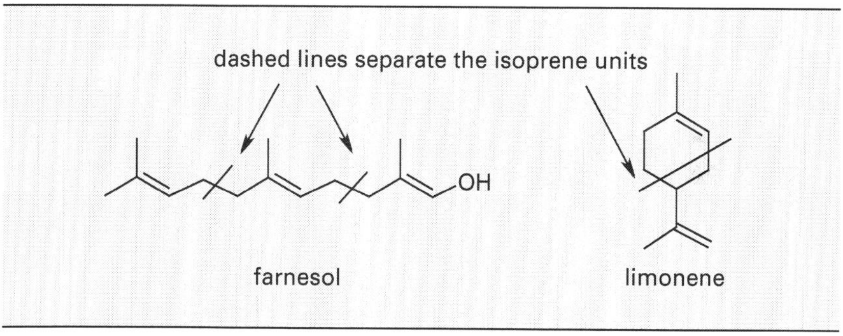

dashed lines separate the isoprene units

OH

farnesol

limonene

Hydrocarbons from biomass. Examples of terpenes constructed from isoprene units (single solid lines denote single bonds between carbon atoms; double solid lines denote double bonds between carbon atoms).
Source: Reference 45.

10

How can we use heat to produce biofuels?

What is thermochemical processing?

In contrast to biochemical processing, which uses enzymes and microorganisms to break down plant polymers and ferment the resulting sugars to fuel molecules, thermochemical processing uses heat and catalysts to transform the chemical energy of biomass into fuels. Thermochemical processing more closely resembles a prairie fire than the ruminations of a cow's stomach: it is hot, fast, and often indiscriminate in what it consumes. Like a fire, it leaves little behind other than ash and sometimes charcoal, consuming even the lignin that microorganisms generally are unable to decompose.

Thermochemical processes were commercially deployed for energy production over two hundred years ago for the production of "town gas" or "manufactured gas," which found wide-spread application until displaced by natural gas. This long history has led to the misperception that thermochemical processes are technologically mature and even obsolete, offering little scope for the kinds of breakthroughs widely forecast for biochemical production of biofuels. By this reasoning, biochemical processing is even more mature since fermentation has been practiced for millennia. A report released in 2008 by the National Science Foundation, the

U.S. Department of Energy, and the American Chemical Society entitled "Breaking the Chemical and Engineering Barriers to Lignocellulosic Biofuels" explains that thermal and catalytic sciences also offer opportunities for breakthroughs in advanced biofuels.[1] Furthermore, the report notes that thermochemical approaches offer distinct advantages over the biochemical approach including the ability to produce a diversity of oxygenated and hydrocarbon fuels (not just alcohols); reaction times that are several orders of magnitude shorter than biological processing; lower catalyst costs; the ability to recycle catalysts; and the fact that thermal systems do not require the sterilization procedures demanded for biological processing. Thermochemical technologies generally convert both carbohydrate and lignin into useful fuel products.

Interest in thermochemical biomass conversion has grown dramatically in the last few years. As recently as 2007 many people were surprised that half of the successful applications to the U.S. DOE's first biorefinery demonstration program incorporated thermochemical processing.[2] Energy companies are supporting thermochemical research at universities and their own research labs. Undoubtedly part of the impetus for developing thermochemical technologies in the U.S. is the Renewable Fuel Standard (RFS) of the Energy Independence and Security Act of 2007, which included a mandate for 16 billion gallons of transportation fuels from cellulosic biomass by 2022. For the last thirty years, the U.S. Department of Energy has been betting on one pathway, cellulosic ethanol, to provide the U.S. with advanced biofuels. There is now widespread realization that multiple pathways will have to be pursued if advanced biofuels are to play an important role in reducing reliance on imported petroleum.[3]

The thermochemical platform offers multiple pathways

to advanced biofuels. These pathways are built upon gasification, fast pyrolysis, and solvolysis, each of which yields distinctive liquid and gaseous intermediate products suitable for upgrading into a wide range of transportation fuels. Thermochemical processing is also the pathway to electricity from biomass, for those that favor battery electric vehicles over liquid transportation fuels.

Gasification to biofuels

Thermal gasification is the conversion of solid, carbon-rich materials at elevated temperatures (typically 750 – 1500°C) and under oxygen-starved conditions into syngas, a flammable gas mixture of carbon monoxide, hydrogen, methane, nitrogen, carbon dioxide, and smaller quantities of hydrocarbons.[4] The technology was introduced in England in 1812 to provide gas lighting from coal and eventually found application in heating, power generation, and production of transportation fuels. This technology was widely adopted in industrialized nations and was employed in the United States as late as the 1950's when interstate pipelines made inexpensive supplies of natural gas available.[5] Some places, such as China, still manufacture gas from coal.

Typical concentrations of gaseous constituents of syngas obtained from an oxygen-blown biomass gasifier.

Gas concentrations (volume percent)					
Hydrogen	Carbon monoxide	Carbon dioxide	Methane	Nitrogen	Heating value (MJ/m)
32	48	15	2	3	10.4

Source: Reference 7.

Gasification is a combination of energy absorbing (endothermic) and energy releasing (exothermic) processes, both of which must occur to give the desired gas product. These processes begin with heating and drying of the biomass, which are endothermic, and bring the biomass to a temperature at which chemical reactions can occur. These are followed by pyrolysis, another endothermic process that decomposes the biomass into charcoal, which is essentially solid carbon, and a mixture of flammable gases and vapors.

Additional reactions are required to convert the charcoal to gas, decompose large gaseous molecules to the carbon monoxide (CO) and hydrogen (H_2) desired in syngas, and to provide thermal energy to drive the various endothermic processes occurring in the gasifier. Some of the charcoal (C) is oxidized by carbon dioxide (CO_2) and water vapor (H_2O) released during pyrolysis to generate additional flammable gases, mostly CO and H_2. In some instances steam is injected into the gasifier to encourage conversion of charcoal to gas. Although these reactions occur readily at the elevated temperatures of gasification, they also are endothermic and will not continue unless the reactor is supplied with heat. Hydrogen formed during gasification can also react with charcoal to form methane (CH_4) and release heat to the gasifying biomass, although this usually does not contribute significantly to the heat demand of the endothermic gasification reactions. An exception is hydrogasification, which is gasification of carbon-rich feedstocks in H_2 gas at relatively low temperature and high pressure, which favors the formation of CH_4. More typically, heat must be added to the gasifier to support the endothermic gasification reactions.

Some systems burn fuel and transfer this energy to the gasification vessel through heated walls or in the form of hot sand. More typically heat is supplied by injecting a relatively

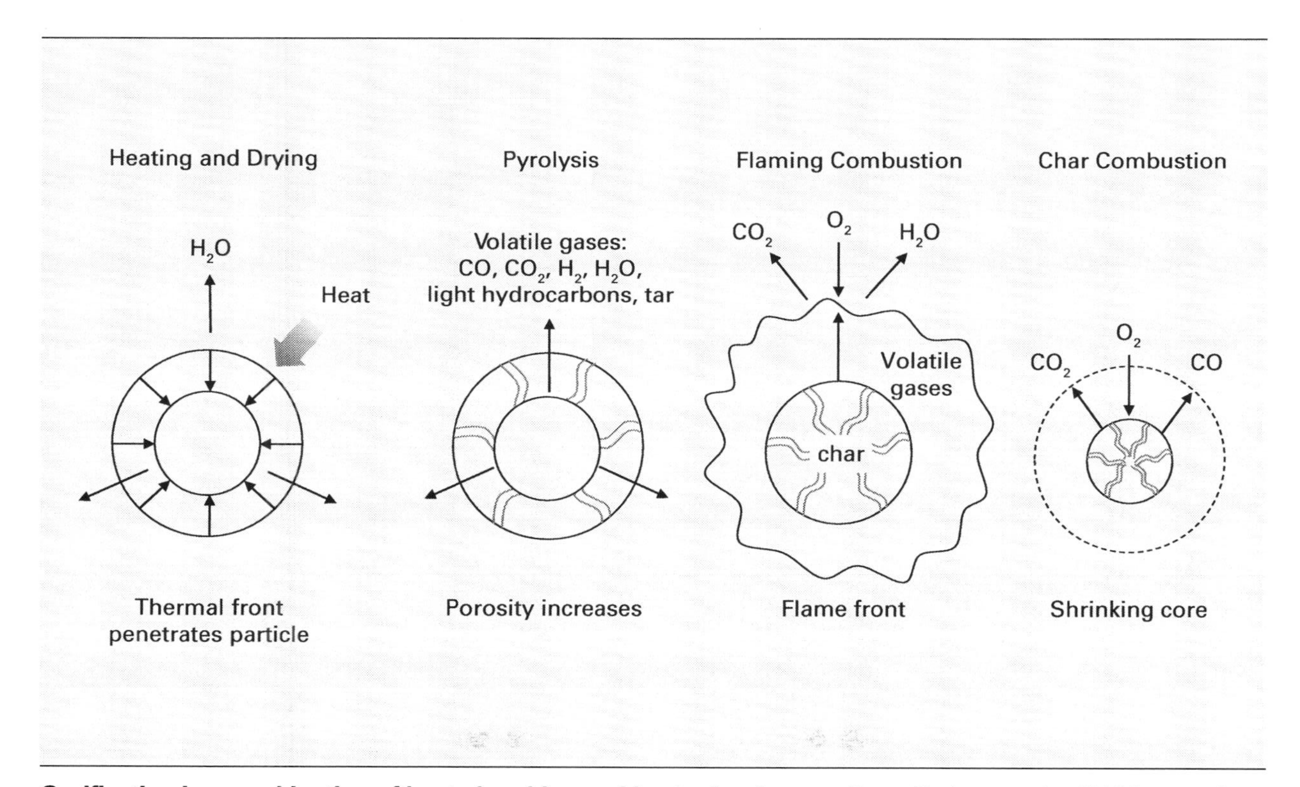

Gasification is a combination of heat absorbing and heat releasing reactions that convert solid biomass into flammable gaseous mixtures. *Source: Reference 4.*

small amount of oxygen or air into the gasifier which burns part of the charcoal or gas product. As explained in Chapter 5, some people mistakenly think such heat addition represents wasted energy. In fact, most of this added energy boosts the energy content of the gaseous product. When the goal is to synthesize fuel molecules, endothermic reactions are preferred over exothermic reactions because they add rather than subtract from the final energy content of the fuel.

Depending upon the design of the gasifier, the gases and vapors released from the biomass react together to determine the final composition of the syngas. If the gasifier is operated at relatively low temperatures and the gases exit the reactor quickly then a wide variety of compounds are found in the gas mixture ranging from very light gases like H_2 and CO to very large molecules commonly referred to as tar, an undesirable product of gasification. As gasifier temperatures and gas residence times in the gasifier increase, reactions are able to more closely approach the condition of chemical equilibrium, where gas compositions are very predictable and usually consist of mostly light gases: CO, H_2, CH_4, CO_2, and H_2O.

Although a variety of gasifier configurations have been developed over the years, they can be broadly classified as non-slagging or slagging gasifiers. Non-slagging gasifiers operate around 850°C, low enough to prevent ash in the biomass from melting. The most common gasifier configuration for non-slagging operation is the fluidized bed, which suspends hot sand particles in an upward moving gas stream. The violent stirring action makes the bed uniform in temperature and composition with the result that gasification occurs simultaneously at all locations in the bed. Biomass injected into this hot medium is converted into syngas, charcoal, and ash particles in a manner of a few minutes. The gas stream contains fine particles of char and ash, tars released from the biomass, and small amounts of sulfur, chlorine, and alkali metals that can contaminate subsequent upgrading steps unless removed. Non-slagging fluidized bed gasifiers have the advantage of relatively simple design and operation with considerable flexibility in the kinds of biomass that can be gasified. A disadvantage is the conversion of carbon in the biomass to syngas may only be ninety percent with the balance appearing as charcoal in the products. It also produces relatively high levels of tar, which can be problematic in

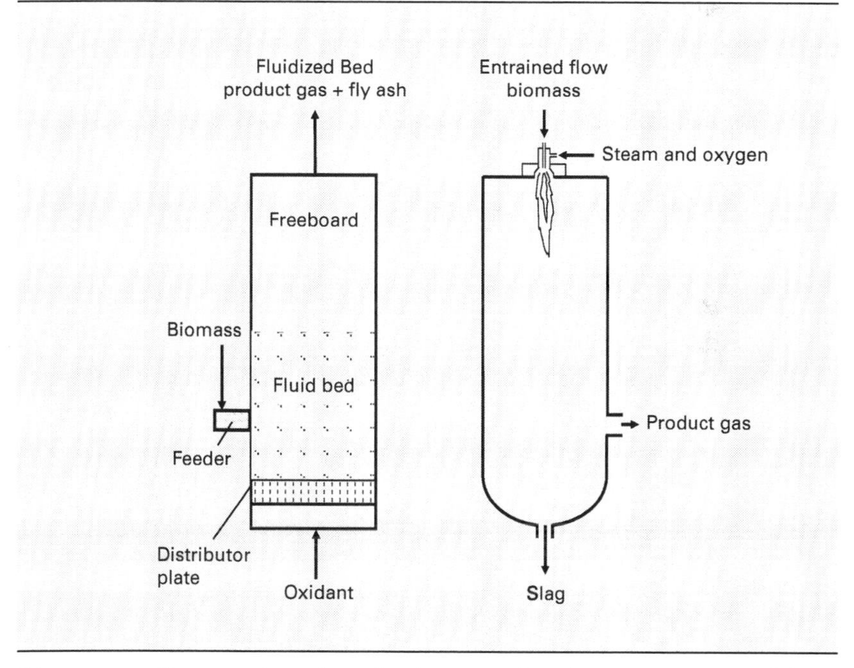

Biomass gasifiers can be categorized as non-slagging, as represented by the fluidized bed gasifier, and slagging, as represented by the entrained flow gasifier. *Source: Author.*

downstream processing to heat, power, or fuels.

Slagging gasifiers operate at temperatures in the range of 1200 - 1500 C, high enough to melt ash released from the biomass. Slagging gasifiers often employ a downdraft, entrained flow configuration, which allows the melted ash (slag) to flow down the walls of the gasifier to a water quench chamber while the syngas exits for subsequent gas cleaning. The slagging entrained flow gasifier has the advantage of carbon conversions often exceeding ninety-nine percent and almost complete elimination of tar in the product gas.

The disadvantage of entrained flow gasifiers is the need for very finely divided fuel, typically to about 0.05 mm diameter, to assure complete gasification of the entrained fuel before it is swept out of the reactor. While coal is sufficiently friable to crush into such fine powders, biomass fibers are relatively resilient and require considerable energy expenditure to achieve size reduction. One possible solution to this problem is to pretreat biomass to yield a fuel form more amenable to feeding into a pressurized gasifier, such as torrefied biomass or slurries of bio-oil and biochar.[6] Torrefaction is a roasting process that heats biomass to 200- 250°C in the absence of oxygen to yield a coal-like product that retains most of the energy content of the original fuel. Unlike biomass, torrefied biomass is readily ground to small diameters with relatively low energy expenditure. Like coal, it can be mixed with water and pumped to high pressure or pneumatically injected under high pressure into an entrained flow gasifier. As described later in this chapter, if biomass is rapidly heated in the absence of oxygen to temperatures around 450°C, it can be converted into liquids (bio-oil) and charcoal (biochar) accounting for up to 93% of the original energy content of the biomass. The bio-oil and biochar can be mixed together to form a pumpable slurry that can be injected under pressure in an entrained flow gasifier.

Conveyance of biomass into an entrained flow gasifier is not the only barrier to its commercialization. The high operating temperatures of entrained flow gasifiers increase their expense and compromise their reliability compared to non-slagging gasifiers. Already commercially developed for coal, slagging entrained flow gasifiers are also being developed for biomass feedstocks.

Syngas can be contaminated by a variety of substances including tar, particulate matter, and compounds of sulfur, chlorine, and nitrogen.[7] Tar can deposit on surfaces in filters, heat exchangers, and engines where they reduce component performance and increase maintenance requirements. Particulate matter, a combination of unreactive mineral matter (ash) and fine charcoal released during pyrolysis, can also clog ducts and foul heat exchanger surfaces and catalyst beds. Sulfur, chlorine, and nitrogen occur as a variety of compounds that will corrode metals surfaces, poison catalysts, and release pollutants into the atmosphere.

The thermodynamic efficiency of gasifiers is strongly dependent on the kind of gasifier and how the product gas is employed.[8] Some high temperature, high-pressure gasifiers are able to convert 90% of the chemical energy of solid fuels into chemical energy of the gas and high temperature heat energy. However, these high efficiencies come at high capital and operating costs. Most biomass gasifiers have conversion efficiencies ranging between 70 and 90%. In many power applications, the hot product gas must be cooled before it is utilized and the high temperature heat of the gas stream is essentially wasted. In this case, "cold gas" efficiency can be as low as 50-70%. Whether the heat removed from the product gas can be recovered for other applications, like steam raising or fuel drying, ultimately determines which of these conversion efficiencies is most meaningful.

Syngas, of course, is not suitable as a transportation fuel despite the occasional efforts to retrofit vehicles with portable gasifiers during World War II and other times of petroleum shortages. Although compressed gases, such as hydrogen and natural gas, are sometimes proposed as energy carriers, syngas is a poor candidate due to its high toxicity arising from its CO content. However, it is a suitable substrate for catalytic and even biocatalytic transformation to a variety of prospective biofuels including hydrogen, methane, ammonia, methanol, ethanol, dimethyl ether, and the same kinds of hydrocarbons found in gasoline and diesel.

The conversion of solid biomass to liquid biofuels or biopower via gasification includes several gas cleaning steps to remove contaminants that might otherwise poison catalysts in fuel synthesis reactors or corrode metal surfaces in power generation equipment. After gasification of the biomass, particulate matter is removed by gas cyclones operated at temperatures high enough to avoid condensation of tarry compounds in the gas. If additional particulate removal is required, this is usually accomplished with ceramic or metal alloy barrier filters able to operate at temperatures required by subsequent upgrading processes. This is followed by tar removal, which is accomplished by either of two methods.[9] One method reacts syngas with steam at high temperature over a bed of catalyst to convert the tar to CO and H_2. Although an attractive approach to tar removal, the catalyst tends to get plugged with a carbon-rich byproduct known as coke. For this reason, scrubbing the gas stream with a fine mist of water or oil is often used as a simple but effective method for removing both tar and particles. Unfortunately, gas scrubbers are not as energy efficient as catalytic beds and produce waste water contaminated with tar, which complicates waste disposal. Sulfur can be removed by a variety of liquid solvents or solid sorbents. The challenge is reducing sulfur concentration in the gas stream to as little as a 50 parts per billion. Nitrogen occurs primarily as ammonia (NH_3) in the syngas. It can be decomposed to harmless nitrogen gas by reacting it with steam over a bed of catalyst in the same manner as tar is decomposed. In fact, tar and ammonia are sometimes decomposed in the same catalytic reactor.[10] Alternatively, ammonia is soluble in slightly acidic solutions and can be scrubbed from syngas by a water spray. Chlorine occurs in the syngas as gaseous hydrochloric acid (HCl), which is also water soluble. Ammonia and chlorine can, in principle, be removed with the same scrubber. At this point the syngas is usually clean enough to be catalytically upgraded or burned in a gas turbine.

Hydrogen has frequently been touted as the future transportation fuel with expectations that it would be renewably generated from solar and wind energy. In fact, the lowest cost hydrogen from renewable resources would likely come from gasification of biomass.[11] Several studies have

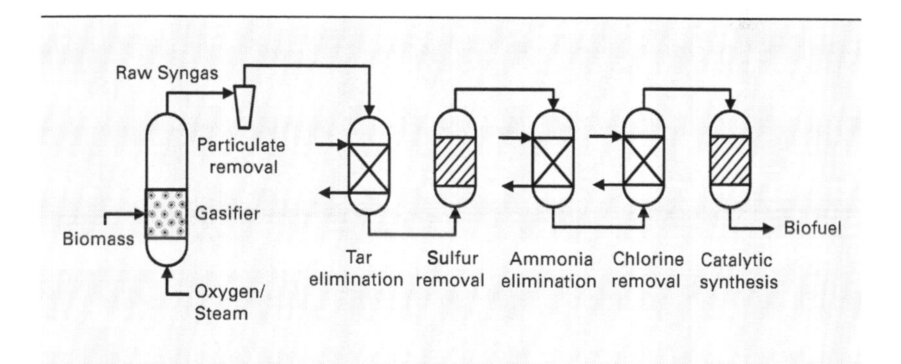

Conversion of solid biomass to liquid biofuels consists of several steps.
Source: Author.

concluded that hydrogen from biomass gasification will cost between $1.07 and $2.98 per gallon of gasoline equivalent using presently available technology.[12, 13, 14, 15] In contrast, production of hydrogen from wind and solar energy using today's best available technology would cost $10.70 and $28.19 per gallon of gasoline equivalent, respectively.[16] Even allowing for advances in electricity generation and hydrogen production and storage technologies in the coming decades, hydrogen costs will still be $4 - $7 per gallon gasoline equivalent, which is 33% to 133% higher than the highest estimates for hydrogen from biomass gasification.

Although syngas is a combination of carbon monoxide and hydrogen, the carbon monoxide can be reacted with steam in the presence of a metal catalyst to generate

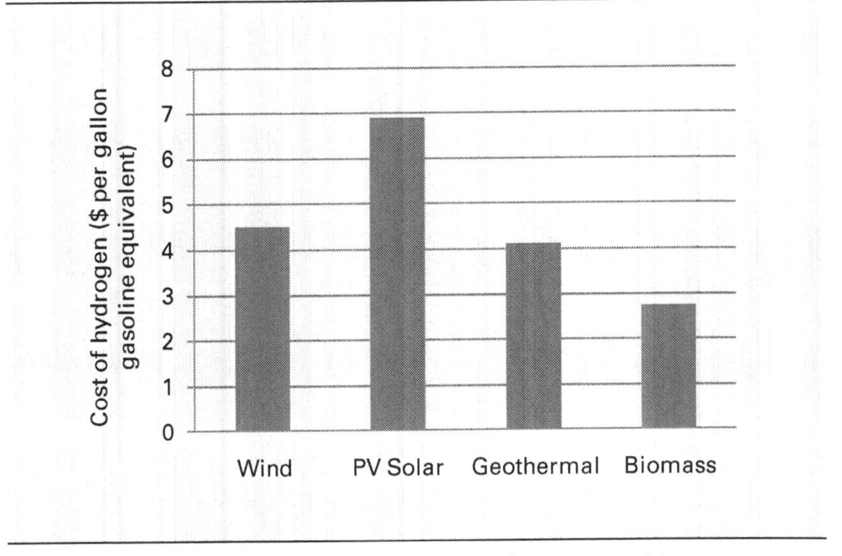

Biomass is the least cost source of hydrogen from renewable resources. Costs are based on projected future available technologies in 2040.
Source: Reference 10.

hydrogen in a slightly exothermic process known as the water-gas shift reaction:

$$CO + H_20 \rightarrow CO_2 + H_2$$

The process resembles electrolysis: water is split into hydrogen atoms and oxygen atoms but instead of forming molecular oxygen (O_2), the oxygen atoms oxidize the CO to form CO_2. This is an example of an exothermic reaction producing an energy carrier (H2) with the carbon monoxide serving as the energy source. Since low temperatures thermodynamically favor exothermic reactions, two reactor stages are employed operating at progressively lower temperatures and with different kinds of catalysts to obtain satisfactory reaction rates.[19] The water-gas shift reaction is already widely employed by ammonia manufacturers, petroleum refiners, and other industries to produce hydrogen from natural gas and other fossil fuels.

Ammonia (NH_3) has been proposed as a propane-like fuel that can be produced from hydrogen and distributed as a compressed liquid using production and distribution technology developed for the nitrogen fertilizer industry.[20] The exothermic reaction of hydrogen and nitrogen to produce ammonia favors high pressures and low temperatures:

$$N_2 + 3H_2 \rightarrow 2\ NH_3$$

Complete reaction would require pressures of 800 atmospheres and reaction temperatures below 200°C. Practical considerations require operation at higher temperatures and lower pressures. The Haber-Bosch process employs an iron-based catalyst at 200 atmospheres and 500°C to achieve a relatively modest 10-20% molar yield of ammonia, but the process is made economically feasible through recycling of unreacted gases and energy recovery.[21] Sulfur readily poisons the iron-based catalyst used to facilitate this reaction. Since the hydrogen is generally obtained through steam reforming and water-gas shift reactions that also employ sulfur-sensitive catalysts, sulfur removal is usually accomplished ahead of hydrogen production. The energy lost in transforming three molecules of hydrogen into two molecules ammonia would appear to be relatively modest: 87% of the chemical energy of the hydrogen appears in the chemical energy of the ammonia. However, the nitrogen for this fuel comes from liquefying air, an energy intensive process. Several energy intensive processes in ammonia manufacture bring the energy tab to 650 kilojoules per mole of ammonia produced.[22] This may be justified when the product is fertilizer, but it does not compare favorably to the 316 kilojoules of energy released when the ammonia is burned as fuel.

Methane can be catalytically generated from syngas reaction over a nickel catalyst by the following exothermic reactions:[26]

$$CO + 3H_2 \rightarrow CH_4 + H_2O$$

$$CO_2 + 4H_2 \rightarrow CH_4 + 2H_2O$$

Synthetic natural gas from coal by this process has been demonstrated at commercial scale although the price of production is too high to be more widely deployed at this time. Of course, costs of production could be reduced if methane was generated at high yields during the gasification process itself, which would eliminate the post-gasification synthesis steps described above. This can be achieved by gasifying at low temperatures and high pressures in the presence of large quantities of hydrogen, which reacts with carbon and carbon monoxide to form methane:[27]

$$C + 2H_2 \rightarrow CH_4$$

$$CO + 3H_2 \rightarrow CH_4 + H_2O$$

Known as hydrogasification, the process requires catalysts and an external source of hydrogen to achieve reasonable reaction rates. The process has not achieved commercial status because suitable low-cost catalysts have yet to be developed.

Methanol (CH_3OH) is commercially produced by reacting syngas over a copper-zinc catalyst at 5-10 MPa and 250°C.[28] Although fossil fuels are currently used to generate syngas for commercial production of methanol, biomass is also a suitable gasification feedstock. The production cost for methanol from biomass is around $1.19 per gallon gasoline equivalence.[14] Although attractive on a cost basis, it has lost favor in recent years because of its toxicity and the fact that it is a precursor for methyl tertiary butyl ether (MTBE), which has been banned as a fuel additive.

Dimethyl ether (CH_3OCH_3) is produced either directly from syngas or indirectly through the dehydration of methanol by reactions at high pressure over catalysts.[29] Alternatively, DME can be dehydrated over a zeolite catalyst to yield gasoline by the energy efficient "methanol-to-gasoline" process developed by Mobil. Methanol-to-gasoline has been commercially demonstrated in New Zealand using natural gas as feedstock. Although yielding high octane gasoline, the products include a high melting point hydrocarbon that can clog fuel injectors.

Ethanol (C_2H_5OH) synthesis is also possible from syngas.[30] Efforts in Germany during World War II to develop alternative motor fuels discovered that iron-based catalysts could yield appreciable quantities of water-soluble alcohols from syngas. Some researchers have advocated the use of "mixed alcohols" as transportation fuels because the product typically contains a mixture of methanol, ethanol, 1-propanol, and 2-propanol. The process was commercialized in Germany between 1935 and 1945 but eventually abandoned because of the increased availability of inexpensive petroleum. Working at pressures of around 50 bar and temperatures in the range of 220-370°C, researchers have developed catalysts with selectivity to alcohols of over 95%, but production of pure ethanol has been elusive. Mixed alcohol synthesis has not been commercialized due to poor product selectivity and low syngas conversion. On a single pass, about 10% of syngas gets converted with most of the product composed of methanol. Recycling methanol aids the production of higher alcohols. Generating revenue from higher alcohols will be important for the successful commercialization of this process. The most recent commercialization effort to turn biomass-derived syngas into alcohol fuels was Range Fuels, which failed in 2011 despite significant public and private financing.[31]

Fischer–Tropsch liquids are synthetic hydrocarbon fuels produced from syngas by the action of metal catalysts at elevated pressures.[32] Depending on the types and quantities of F-T products desired, either low (200–240°C) or high (300–350°C) temperature synthesis is used with either an iron (Fe) or cobalt catalyst (Co). Additional processing of the F-T products yields diesel or gasoline. Fischer-Tropsch technology was extensively developed and commercialized in Germany during World War II when it was denied access to petroleum-rich regions of the world. Likewise, when South Africa faced a world oil embargo during its era of apartheid, it employed Fischer-Tropsch technology to sustain its national economy from indigenous sources of coal. Fischer-Tropsch catalysts are readily poisoned by sulfur, nitrogen, and chlorine at concentrations below one part per million. Thus, removal of contaminants ahead of the synthesis reactors is an extremely important and expensive part of a Fischer-Tropsch synfuels plant. With biomass as feedstock, Fischer-Tropsch fuels have been variously estimated to cost between $2-$5 per gallon gasoline equivalent although recent analysis has suggested costs will be closer to the higher end of this range.[55] A number of companies have plans to commercialize biomass to Fischer-Tropsch fuels including Rentech and Syntroleum in the United States.

Syngas can also be biocatalytically upgraded to biofuels in a process known as syngas fermentation. A number of microorganisms are able to utilize carbon monoxide, carbon dioxide, and hydrogen as substrates for growth and production of hydrogen, alcohols, carboxylic acids, and esters.[56] Syngas fermentation has several advantages compared to catalytic upgrading of syngas.[57] Most catalysts used in the petrochemical industry are readily poisoned by sulfur-bearing gases whereas syngas-consuming anaerobes are sulfur tolerant. Unlike metal catalysts, biocatalysts do not require high pressure or specific CO/H_2 stoichiometries. Gas-phase catalysts typically employ temperatures of several hundreds of degrees Centigrade and at least 10 atmospheres of pressure whereas syngas fermentation proceeds at near ambient conditions.

Several barriers must be overcome before syngas fermentation is commercially viable including growth rate and production of microorganisms, product inhibition by acids and alcohols, and limitations of syngas mass transfer.[26] The idea of harnessing microorganisms as syngas catalysts was proposed in the early 1980's but received relatively little attention until recently. Companies developing syngas fermentation include Bioengineering Resources, Inc.; Coskata; INEOS Bio; and LanzaTech. Very little economic data has been published on this process, although Coskata optimistically estimates production costs (exclusive of capital recovery) could be as low as $1.00 per gallon of gasoline equivalent product.[31]

Fast pyrolysis to biofuels

Fast pyrolysis is the rapid thermal decomposition of biomass in the absence of oxygen to produce liquid, gas, and char.[32] Fast pyrolysis occurs at lower temperatures (about 500°C) and shorter residence times (a few seconds) compared to gasification. Whereas the primary product of gasification is syngas, the primary product of fast pyrolysis is liquid, with mass yields as high as 75% of the biomass input. The liquid, commonly called bio-oil, contains 15 to 20% water by weight with the balance made up of a complex mixture of oxygenated organic compounds including carboxylic acids, alcohols, aldehydes, ketones, esters, saccharides, and phenolic monomers and oligomers.[33] Although these compounds are not generally suitable as transportation fuel, they can be upgraded for this purpose, as subsequently described. Assuming conversion of 72% of the biomass feedstock to liquid on a weight basis, yields of pyrolysis oil are about 135 gallons per ton of biomass.

Rapid heating of biomass followed by rapid cooling of products is essential if high yields of liquids are to be

Bio-oil is liquid biomass with similar elemental composition to the solid biomass from which it is derived. *Source: Iowa State University.*

produced instead of non-condensable gases and charcoal. A number of pyrolysis reactors have been developed to accomplish this process. The most thoroughly investigated design is the fluidized bed pyrolyzer which exploits the turbulent environment produced when gas is injected upward through a bed of sand. Lignocellulosic feedstock, such as wood or agricultural residues, is milled to a few millimeters size to promote rapid reaction. The particles are conveyed into the fluidized bed reactor where they are rapidly heated and converted into condensable vapors, liquid aerosols, non-condensable gases, and charcoal.[34] These products are transported out of the reactor into a cyclone operating above the condensation point of pyrolysis vapors where the charcoal is removed. Vapors and gases are transported to a quench vessel where a spray of pyrolysis liquid cools vapors sufficiently for them to condense. The non-condensable gases, which include flammable carbon monoxide, hydrogen, and methane, are burned in air to provide heat for the pyrolysis reactor.

Dynamotive, Ensyn Technologies, and Avello Bioenergy have developed fluidized bed pyrolyzers for the production of boiler fuel and value-added products from bio-oil. Other kinds of reactors that have been commercialized for bio-oil production include the rotating cone pyrolyzer developed by BTG, Inc. in the Netherlands and screw pyrolyzers offered by several companies including ABRI, Inc. of Canada

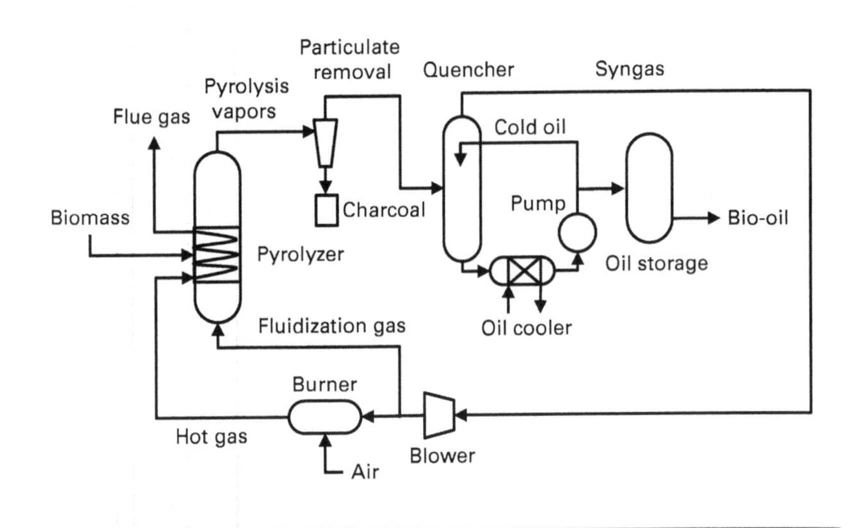

Bio-oil production system based on a fluidized bed reactor.
Source: Author.

and ROI of the United States.

The relative simplicity of generating bio-oil, the prospect for its distributed generation, and the convenience of storing a liquid intermediate instead of a gaseous intermediate before upgrading is responsible for the growing interest in fast pyrolysis. However, raw bio-oil has several prominent shortcomings.[35] The high carboxylic acid content makes it extremely corrosive. Poor separation of particulate matter and condensable vapors may result in char and mineral matter in the oil, which can clog fuel nozzles and catalyze reactions that destabilize the bio-oil mixture over time. Its heating value of 17 to 20 MJ/kg is only half that of petroleum-based heating oil. Bio-oil can be used as boiler fuel or fired in gas turbines, but it is not suitable as transportation fuel. Thus, bio-oil must be upgraded to liquid products more useful as transportation fuels.

Bio-oil is more complex than the sugar streams fermented in an ethanol plant or the syngas used in Fischer-Tropsch synthesis. The three important components of bio-oil from the perspective of upgrading are water, carbohydrate-derived compounds, and lignin-derived compounds. The water derives from both moisture in the original biomass and "produced water" from chemical reactions occurring during pyrolysis. Water can account for 15-20% of the weight of bio-oil.

Carbohydrate-derived compounds include carboxylic acids (such as acetic acid), aldehydes (such as furfural), ketones (such as acetone), and even sugars. Interestingly, pure cellulose decomposes mostly to a dehydrated sugar known as levoglucosan,[36] which can be readily converted to glucose, the same simple sugar produced from starch for ethanol production. However, naturally occurring potassium and calcium in biomass, which are important nutrients for plant growth, interfere with this thermal pathway to sugars, resulting in the production of less desirable acids, ketones, and aldehydes.[37] Most of these compounds are sufficiently water soluble to end up in the aqueous phase of the bio-oil.

The lignin-derived compounds are responsible for much of the complex composition of bio-oil. Virgin lignin, as found in the original biomass, is itself a complex polymeric material with a distinctive "chicken-wire" structure based upon phenol, a ring-shaped molecule constructed of six carbon atoms with a hydroxyl (OH) side group. Considering that lignin is both the "glue" that holds together cellulose strands in plant fibers and an anti-microbial agent that protects plant fibers from pests, it is not surprising that phenol has commercial and historical applications in the production of adhesives and antiseptics.[38]

When lignin is heated, it proceeds through stages of

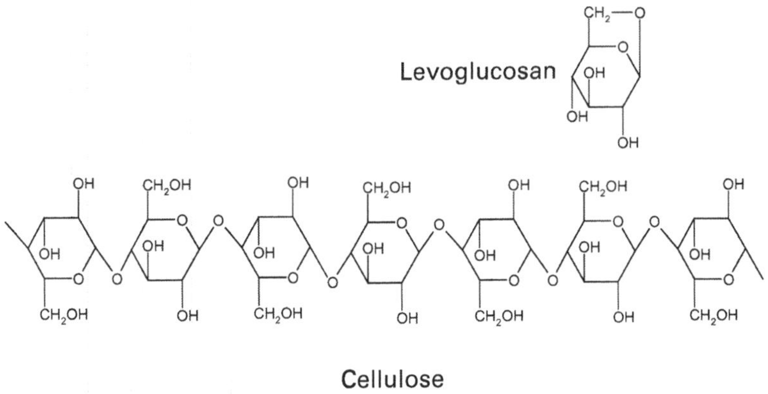

Levoglucosan

Cellulose

Cellulose would thermally decompose to the dehydrated sugar levoglucosan except that naturally occurring potassium and calcium in biomass catalyze its decomposition to smaller molecules. *Source: Author.*

softening, melting and ultimately decomposition into smaller molecules, a process that is poorly understood. However, just as cutting up a large sheet of chicken wire yields progressively smaller networks of connected rings until only single rings of metal wire with small side chains remain, the decomposition of the lignin polymer first yields large molecular fragments known as lignin oligomers, which are non-volatile. Further chain breaking eventually yields single rings of six carbon atoms (monomers) with short side chains containing various combinations of carbon, hydrogen, and oxygen, a large class of chemicals known as substituted phenols or phenolic compounds. These short side chains impart a wide variety of physical and chemical properties to substituted phenols including acidity, boiling point, solubility, and reactivity.

The phenolic monomers have sufficiently low boiling points that they can evaporate as they are formed and escape the pyrolyzer and subsequently condense in the bio-oil recovery system. The lignin oligomers, though, have very high boiling points and do not readily evaporate. Some researchers think that the lignin oligomers are blown out of the pyrolyzer as liquid droplets.[39] However, there is increasing evidence that if these non-volatile oligomers are not able to escape through decomposition to monomers their fate is to dehydrate to charcoal.[40]

Interestingly, though, analysis of bio-oil reveals relatively few phenolic monomers compared to the amount of lignin in the bio-oil. Instead, phenolic monomers appear to recombine to form oligomers that superficially resemble the original lignin.[41] These water insoluble oligomers are known as pyrolytic lignin. They form emulsions with the aqueous phase of bio-oil, which can exist as stable mixtures for several months, depending upon how they are produced.

The pyrolytic lignin can be readily precipitated from the aqueous phase by addition of water to bio-oil, which has advantages in upgrading the bio-oil to transportation fuels.

Successful upgrading of bio-oil may require different approaches for the aqueous phase, which contains most of the carbohydrate-derived compounds and the water insoluble phase, which contains most of the lignin-derived compounds. The aqueous phase consists of many molecules with carbon chains too short for synthetic gasoline or diesel while the insoluble phase is primarily molecules that are too large to serve as fuel.

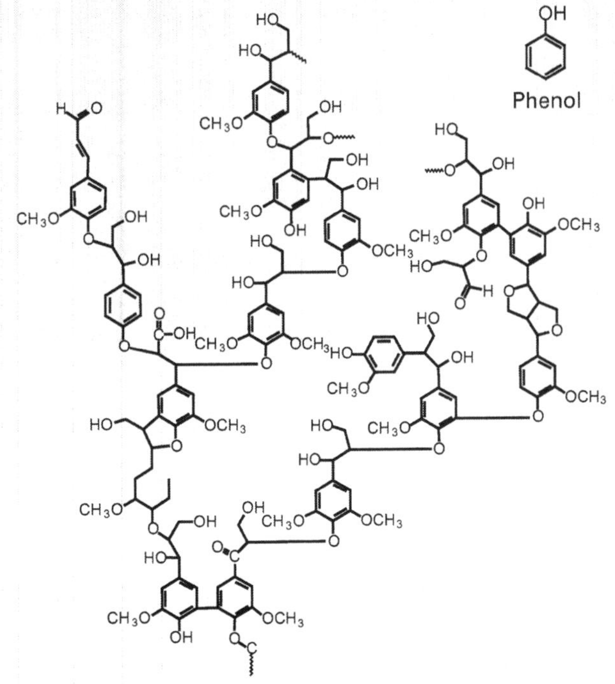

Phenol

Phenol is the building block of lignin. *Source: Reference 42.*

Two methods for upgrading the aqueous phase are being contemplated. The first, called steam reforming, reacts the aqueous phase at elevated temperatures over a catalytic bed to form hydrogen and carbon dioxide. This hydrogen is used to upgrade the insoluble fraction of bio-oil, as subsequently described. The process is technically feasible but actually produces more hydrogen than is needed to upgrade the insoluble fraction of bio-oil. Accordingly, steam reforming all of the aqueous phase would reduce theoretical yields of synthetic hydrocarbon fuels from bio-oil. Alternatively, all or part of the aqueous phase is passed over a bed of catalyst where larger molecules are synthesized by a process known as aqueous phase processing.[43]

Upgrading the insoluble fraction of bio-oil to hydrocarbons can be accomplished by hydrotreating and hydrocracking, processes which were originally developed to convert petroleum into gasoline and diesel. Hydrotreating is the reaction of organic compounds with hydrogen over a catalyst at high pressure to remove sulfur, nitrogen, oxygen, and other contaminants. Whereas oxygen is a relatively minor contaminant in petroleum, it represents a major constituent in bio-oil, which may require catalyst formulations optimized for deoxygenation of bio-oil. Hydrocracking is the reaction of hydrogen with organic compounds to break long-chain molecules into lower molecular weight compounds. Although fast pyrolysis breaks down large plant polymers, a number of large molecular fragments remain especially pyrolytic lignin. Hydrocracking breaks these oligomers into molecules closer in size to those needed for the production of transportation fuels. Hydrotreating and hydrocracking of lignin oligomers and phenolic compounds derived from lignin will generate both straight-chain hydrocarbons and aromatic compounds, which are built from the same six-carbon ring found in phenolic compounds. Aromatic hydrocarbons are an important component of some gasoline blends and most aviation fuels. A report by UOP suggests that gasoline from bio-oil will cost $1.70 - 1.90 per gallon,[44] while researchers at Iowa State University calculate a minimum cost of $2.11 per gallon.[45]

Some researchers are attempting to add catalysts to the pyrolysis reactor to directly yield hydrocarbons. Similar to the process of fluidized catalytic cracking used in the petroleum industry, the process occurs at atmospheric pressure over acidic zeolites. Yields of up to 17% C_5-C_{10} hydrocarbons have been reported in a study of upgrading of pyrolytic liquids from woody biomass.[46,47] Although superior to conventional bio-oil, this product still needs refining to gasoline or diesel fuel.

Although bio-oil is not usually considered a substrate for fermentation, the dehydrated sugars released during the decomposition of cellulose and hemicellulose can be hydrolyzed to fermentable sugars. The amount of sugars formed can be enhanced by either removing potassium and calcium from biomass[48] or otherwise deactivating the catalytic activity of these metals[49] prior to pyrolysis. Limited technoeconomic analysis of the process suggests that a bio-oil fermentation plant could produce ethanol at costs competitive with cellulosic ethanol derived from acid or enzymatic hydrolysis.[50]

Purely catalytic approaches to converting sugars to fuel molecules are also possible. Sugars that exist as five-member rings like the five-carbon sugar xylose or the six-carbon sugar fructose are readily dehydrated to the five-member rings of furan compounds.[51] Furans are colorless, water-insoluble, flammable liquids with volatility comparable to hydrocarbons of similar molecular weight. Some kinds of furans have heating values and octane numbers comparable to gasoline making them potential transportation fuel.[52] Catalysts can improve yields by making furan-producing pathways more selective among the large number of competing reactions that can occur during pyrolysis of biomass. 2,5 dimethyl furan in particular has received recent interest because new catalytic synthesis routes from sugars have been developed.[53,54] Neither the fuel properties nor the toxicity of these compounds has been much studied raising questions as to their ultimate practicality as transportation fuel.

A more promising approach reacts monomeric sugar or sugar-derived compounds in the presence of heterogeneous catalysts at 200-260°C and 10-50 atmospheres of pressure to produce alkanes, the same hydrocarbons found in gasoline.[55] Catalytic conversion of sugars would have several advantages over fermentation, including higher throughputs, ready conversion of a wide range of sugars, and the immiscible hydrocarbon products could be recovered without the expensive distillations required in ethanol plants. Of course, the process would require large quantities of inexpensive sugars, the same barrier to continued growth of the first-generation ethanol industry. Thermal depolymerization of

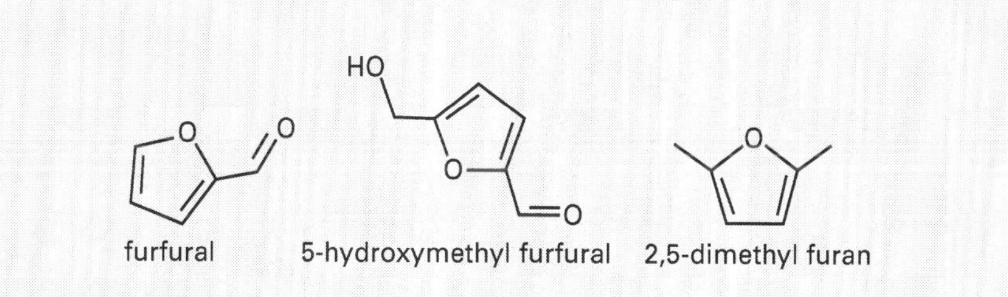

furfural 5-hydroxymethyl furfural 2,5-dimethyl furan

Furans relevant to the production of transportation fuels by thermochemical processing of sugars. *Source: Reference 43*

cellulose through fast pyrolysis or hydrothermal processing could potentially supply these sugars as could enzymatic hydrolysis being developed for cellulosic ethanol production. Virent Energy Systems, which is commercially developing aqueous phase processing for the production of hydrocarbons from sugar feedstocks, reports that the process is economically attractive at petroleum prices above $60 per barrel.

Solvolysis

Solvolysis is pyrolysis in the presence of a solvent.[56] Sometimes the process is known as direct liquefaction although this overlooks the prospects for also fractionating or gasifying biomass via solvolysis. A variety of solvents can be employed, their solution chemistry and boiling points determining the temperature and pressures at which the processes operate. Because water is frequently a major constituent of biomass feedstocks, it is frequently employed as a solvent, in which case solvolysis is referred to as hydrothermal processing (HTP).

Hydrothermal processing of wet biomass can produce carbohydrate, liquid hydrocarbons, or gaseous products depending upon the reaction conditions, often without catalysts.[57] As reaction temperature increases, higher pressures are required to prevent boiling of the water in the wet biomass. At temperatures around 100°C, extraction of high value plant chemicals such as resins, fats, phenolics, and phytosterols is possible. At 200°C and 20 atmospheres of pressure, fibrous biomass undergoes a fractionation process to yield cellulose, lignin, and hemicellulose hydrolysis products such as xylose. Further hydrothermal processing can hydrolyze the cellulose to glucose. Renmatix is among companies developing this thermochemical fractionation technology.

At 300 – 350°C and 120 – 180 atmosphere pressure, biomass undergoes more extensive chemical reactions yielding a hydrocarbon-rich liquid known as biocrude. Although superficially resembling bio-oil, it has lower oxygen content and is less miscible in water, making it more amenable to hydrotreating. A number of companies have explored the commercial development of this technology including Biofuel B. V., Catchlight Energy, Changing World Technologies,

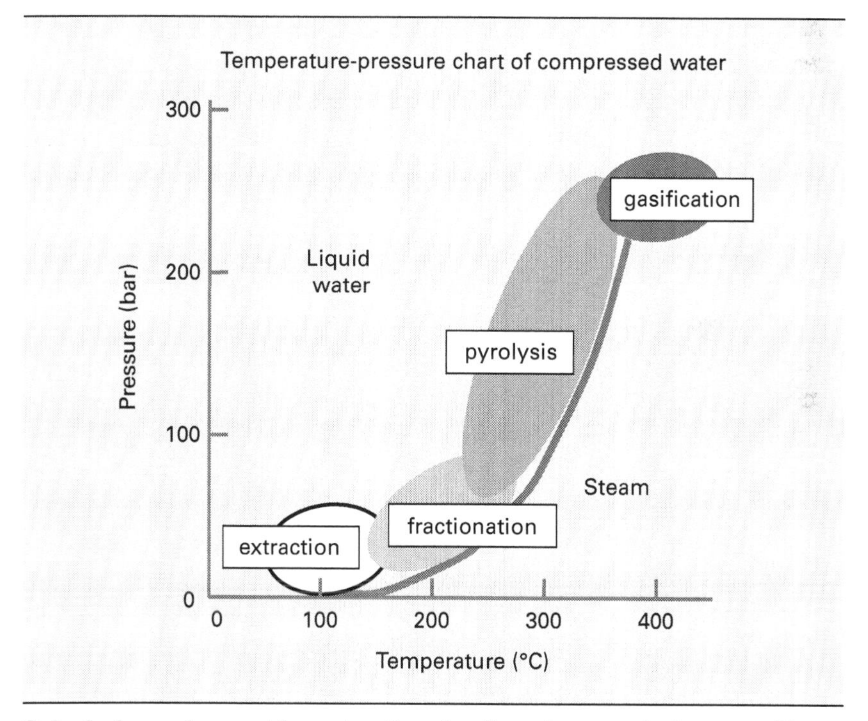

Solvolysis can be used for extraction, fractionation, pyrolysis, or gasification of biomass depending upon the solvent employed and the temperature-pressure region of its operation. *Source: Author.*

EnerTech Environmental, Inc., Shell, and TNO.

At 600 - 650°C and 300 atmospheres pressure the primary reaction product is gas, including a significant fraction of methane.[58] Continuous feeding of biomass slurries into high pressure reactors and efficient energy integration represent engineering challenges that must be overcome before HTP results in a commercially viable technology.

Biopower

As described in Chapter 5, electric propulsion is an attractive alternative to liquid transportation fuels if inexpensive batteries can store enough energy to approach the driving range of vehicles running on liquid fuel. Considering that the much vaunted lithium-ion battery, the most recent advance over traditional lead-acid batteries, has only moved us 2% closer to this goal does not give much encouragement that battery electric vehicles will overtake the internal combustion engine or jet engine for long-distance transportation. Nevertheless, renewably-generated electricity has more immediate prospects for short-distance commuting and delivery service, electric trains, and stationary power applications. The chemical energy of biomass can be converted to electric power for less than $0.10 per kilowatt hour, which is comparable to wind power and one-half to one-fifth the cost of solar photovoltaic power.[59] Unlike intermittent wind and solar power sources, biopower is dispatchable on demand. Biomass can be converted to electric power through a variety of power cycles. Most prominently among these are the Rankine cycle and two variations of combined cycles that employ the Rankine cycle in combination with the Brayton cycle or fuel cells.

The Rankine cycle involves the direct combustion of

fuel to raise pressurized steam that is expanded through a turbine to produce electricity.[60] Steam power plants are the basis of much of the electric power generation capacity in the world. The reason for the Rankine cycle's preeminence has been its ability to directly fire coal and other inexpensive solid fuels. Constructed at scales of several hundred megawatts, the modern steam power plant can convert as much as 45% of chemical energy in fuel to electricity at a cost of about \$0.05 – \$0.10/kilowatt-hour. However, on a volume basis most forms of biomass have less than half the heat content of coal,[61] which makes it difficult to supply a large power plant with sufficient biomass. For this reason biomass power plants are envisioned to be less than fifty megawatts in size, which raises a fundamental difficulty with the Rankine cycle. Such small power plants are less efficient than larger power plants, consuming as much as 20% more energy per unit of electricity produced than large, coal-fired

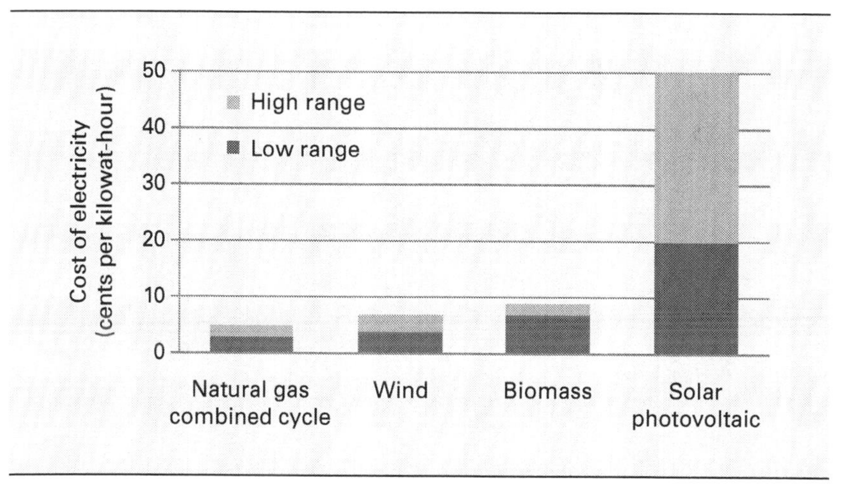

Biopower is comparably priced to wind power with the additional advantage of being dispatchable. *Source: Reference 49.*

power plants.[62] Since biomass is more expensive than coal, alternatives to small-scale Rankine cycles will be required if biopower is to play a significant role in electric power generation.

One approach to solving this limitation of Rankine power cycles is to increase the density of biomass sufficiently for its economical transport to large, central power plants where it can displace a significant fraction of coal presently used as boiler fuel. Torrefaction and fast pyrolysis discussed earlier in this chapter in connection to providing feedstock for pressurized gasifiers could serve this purpose. Torrefied biomass, which resembles coal in many respects, could be pelletized to a high density material that could be handled and transported much like crushed coal. Fast pyrolysis produces a high density, pumpable bio-oil, which could be handled like fuel oil. These could be directly substituted for coal with relatively less difficulty than unprocessed biomass.

Another approach is to substitute power cycles with better prospects for high efficiency electricity generation at modest scales. The combined cycle has been developed in recognition that waste heat from one power cycle can be used to drive a second power cycle.[63] Combined cycles would be unnecessary if a single heat engine could be built to operate between the temperature extremes of burning fuel and the ambient environment. However, temperature and pressure limitations on materials of construction have prevented this realization. Combined cycles employ a topping cycle operating at high temperatures and a bottoming cycle operating on the rejected heat from the topping cycle. Most commonly, combined cycle power plants employ a gas turbine for the topping cycle and a steam turbine for the bottoming cycle, achieving overall efficiencies of 50% or higher. Power plants based on high temperature fuel cells are sometimes integrated with both a gas turbine topping cycle and a steam turbine bottom cycling to further improve efficiency.

The Brayton cycle produces electric power by expanding hot gas through a turbine.[64] Directly firing biomass to generate the hot gas stream has proved impractical because particulate matter and corrosive compounds carried with the gas stream damage the gas turbine. More promising is gasification or fast pyrolysis of biomass to generate syngas or bio-oil that can be cleaned before firing in the gas turbine. Brayton cycles are one of the most promising technologies for bioenergy because of the relative ease of plant construction, cost-effectiveness in a wide range of sizes (from tens of kilowatts to hundreds of megawatts), and the potential

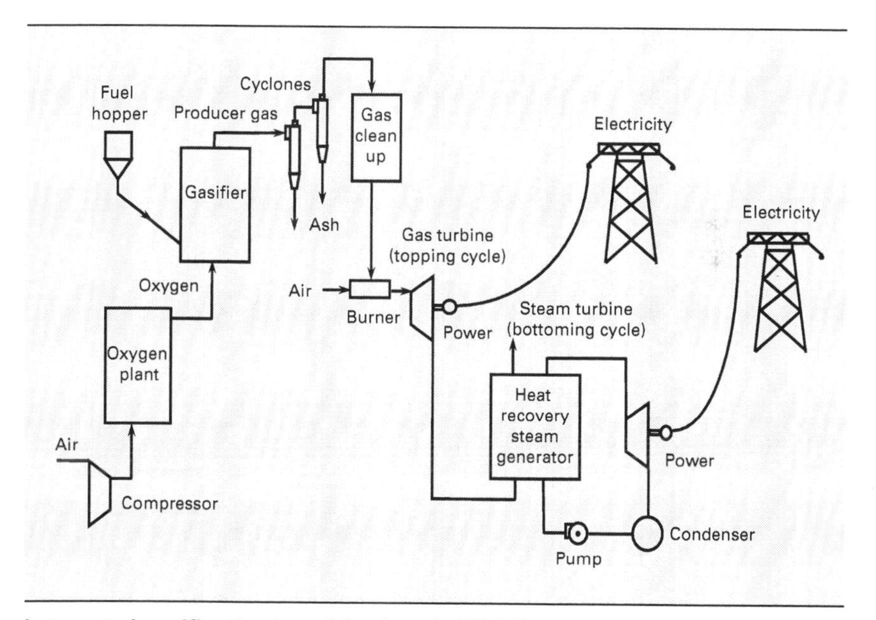

Integrated gasification/combined cycle (IGCC) power plants will be able to convert biomass to electricity at high thermal efficiencies. *Source: Author.*

for high thermodynamic efficiencies when employed in advanced cycles, as subsequently described.

Fuel cells directly convert chemical energy into work, thus bypassing Carnot limits for heat engines.[65] This does not imply that fuel cells can convert 100% of chemical enthalpy of fuel into work. In practice, irreversibilities limit fuel cell conversion efficiencies to 35 – 60%, depending upon the fuel cell design. Thus, fuel cells can produce significantly more work from a given amount of fuel than can heat engines. However, carbonaceous fuels must first be reformed to hydrogen before they are suitable for use in fuel cells. The energy loses associated with fuel reforming must be included when determining the overall fuel-to-electricity conversion efficiency of a fuel cell. Proton exchange membrane (PEM) fuel cells operate at relatively low temperatures (65°C) that are suitable for automotive applications, but high costs associated with generating hydrogen without impurities have limited its commercial application. High temperature fuel cells are most favored for stationary power generation because of opportunities for heat recovery.

11

What is the future of fuels?

What will be the primary energy source and energy carrier for transportation fuels?

One thing is certain about our future sources of primary energy: we will obtain them from diverse sources including fossil fuels, nuclear energy, and renewable energy. This does not mean the mix will resemble our current supplies of energy. Growing demand for energy in the developing world and environmental regulations, most prominently restrictions on greenhouse gas (GHG) emissions, could dramatically change the composition of our energy supply.

Among the fossil energy sources, petroleum will inevitably represent a smaller fraction of the overall mix of primary energy sources as the developing world increases its demand for energy. If price and supply were the only considerations in energy decisions, then coal would return to the prominence it held in the nineteenth century and early twentieth century. In terms of environmental performance, though, coal is among the least attractive energy options. In addition to being a source of acid rain precursors, fine particulate matter, and mercury, it is among the most carbon intensive fuels. Considering the constraints on greenhouse gas emissions that exist in many countries, coal is unlikely to prosper until cost-effective means of capturing and storing carbon dioxide are devised. Whereas the projected price of carbon under taxation and cap-and-trade schemes is in the range of $20-$30 per ton of carbon dioxide equivalent, capturing carbon from coal-fired power plants is expected to cost $50-75 per ton[1, 2, 3] and likely significantly more for the first plants built.

The recent deployment of technology to recover "unconventional" natural gas from shale deposits and "tight sands" is improving the prospects for this relatively clean-burning and low carbon fuel.[4] It is more attractive as a replacement for coal in electric power generation than as a replacement for petroleum in the production of transportation fuels. Retrofitting a coal-fired power plant for natural gas is straightforward and cost-effective while retrofitting automobiles and fueling stations to use natural gas would be cumbersome and expensive.

Nuclear energy is attractive for its low GHG emissions and the large amount of fossil fuels it could potentially displace. Although originally developed to provide power to the electric grid, nuclear energy could power the nation's transportation system in the form of electricity for battery electric vehicles or hydrogen generated through electrolysis for fuel cell vehicles. Both electric- and hydrogen-fueled vehicles have been demonstrated and the former commercialized, but storing these forms of energy is expensive and impractical at present for long-range vehicles. Possibly even more daunting are the infrastructure changes that would have to be made to the world's transportation systems to accept these fuels on a large scale. The 2011 meltdown at the Fukushima nuclear plant in Japan has caused many countries to turn against nuclear power, with Germany phasing out its existing plants ahead of schedule and Italy scrapping its plans to build new plants. Even if public opinion and

U.S. policy become more welcoming toward nuclear energy, it is unlikely to impact the transportation system for many decades to come.

The largest technically recoverable renewable energy resources in the United States are solar and geothermal energy. Solar energy encompasses several manifestations of sunlight interacting with the biosphere and man-made systems that can be classified as either "direct solar insolation" converted to electricity (solar photovoltaic devices and solar thermal cycles) or "indirect solar capture" (wind, waves, and biomass). Of these different forms of renewable energy, the most widely distributed geographically are direct solar insolation, wind, and biomass. Thus, the grand challenge of renewable energy is ultimately discovering new ways to harness solar energy for generation of electric power and production of fuels. Generation of electricity from wind is already economical. Generation of electricity through photon-induced charge separation in photovoltaic devices or through concentrating sunlight to run a thermal power cycle still requires technological advances for wide spread adoption. Nature has already mastered solar energy for the production of fuels: photosynthesis accomplishes both electric charge separation and chemical synthesis of energy-rich carbohydrates and lipids from carbon dioxide (CO_2) and water (H_2O). The U.S. Department of Energy (U.S. DOE) has aspirations of not only emulating photosynthesis, but also doing it more efficiently than Nature.[5] In the meantime, Nature is capturing solar energy in the form of biomass at a rate six times faster than mankind consumes energy. Biomass is among the most cost-effective forms of renewable energy available today, whether for generation of electric power or production of transportation fuels.

Whether cellulose, lipids, or even sugar will dominate

the production of future biofuels is the subject of much conjecture at present. For over thirty years most biomass advocates assumed that cellulose would be the primary if not only feedstock for advanced biofuels. After all, cellulose is the most plentiful form of biomass on the planet. The energetics for producing this highly oxygenated plant polymer are more favorable than for highly reduced lipids. The yields of lipids in terms of units of energy per unit of land area for most cultivated plants are far less than yields of cellulose. As for sugar, the assumption was that it would not be a cost-effective feedstock for large-scale fuels production.

This conventional wisdom has recently been turned on its head. In 2008, several airlines formed a "Sustainable Aviation Fuel Users Group" with a strong focus on producing fuels from lipids.[6] In July 2009, Exxon Mobil, which had until that time avoided any kind of investment in renewable energy technologies, announced a $600 million research effort in microalgae.[7] The U. S. DOE, which closed its microalgae programs in 1996 after almost two decades of research,[8] is once again investing tens of millions of dollars on "algal biofuels."[9]

The aviation industry, faced with looming mandates to reduce the carbon footprint of air travel, have little choice but to synthesize hydrocarbon fuels from lipid-rich biomass (in the longer term carbon capture and sequestration technology may open the way for manufacturing relatively low-carbon drop-in fuels from coal although these will always be inferior to biofuels in terms of carbon footprint). Ethanol, regardless of the source of carbohydrate, is a poor candidate for aviation fuel because of its relatively low energy density and affinity for water. Although companies like Amyris aspire to ferment sugars into hydrocarbons known as terpenes that can be upgraded into aviation fuels, the prospects for

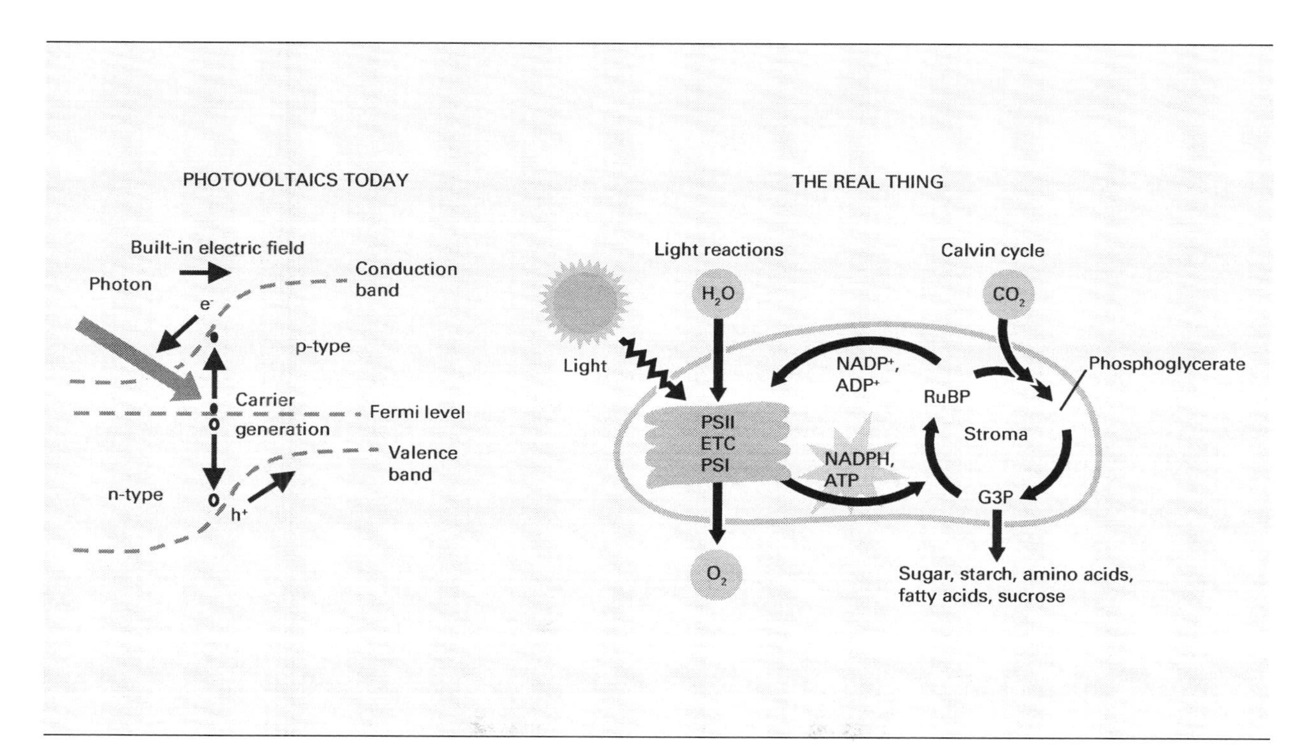

Artificial photosynthesis will require systems based on multiple electron transfer including a donor molecule that can absorb visible light and release many electrons and a receiver molecule capable of accepting and storing those electrons. *Source: Author.*

near-term commercialization appear to be receding. By the end of 2008, commercial aviation test flights had begun on "second generation" biofuels produced by hydrotreating lipids from "non-food biomass."[10] In this case, the biomass was jatropha although the industry hopes eventually to utilize microalgae, which has the potential to produce many times more lipid on a unit of land surface than can conventional lipid crops like soybeans or even emerging crops like jatropha.

Policy makers appear to be moving toward the same conclusion with respect to ground transportation: alternative fuels will be adopted more quickly and at dramatically lower capital investment if they are infrastructure compatible. These so-called "drop-in" fuels would also solve the "ethanol blend wall," which occurred when the production of ethanol exceeds the amount required to produce E-10, a blend of 10% ethanol with gasoline.[11] It would appear that the future energy carrier for transportation is the present energy carrier—hydrocarbons. The difference will be the synthesis of these hydrocarbons from renewable sources instead of their refining from petroleum. This conclusion does not dismiss the possibility that electricity will eventually play a significant role in ground-based transportation systems, but batteries to store electricity are unlikely to approach the energy densities of hydrocarbon fuels. Electricity will become attractive for commuter and local delivery vehicles, high-speed electric rail service, and hybrid-electric vehicles. Fuels based on liquid hydrocarbons will dominate air travel and inter-urban transportation for many decades hence.

Hydrocarbon fuels can be synthesized from either lipids or carbohydrates. By virtue of their lower oxygen content lipids can be converted more readily to hydrocarbons than can carbohydrates. On the other hand, carbohydrates are easier to grow in large quantities than are lipids (as evidenced by the prevalence of hay fields and woodlots over algal lagoons in agriculture). The question is whether growing lots of lipid-rich biomass will prove easier than chemically transforming carbohydrate into hydrocarbons. The commercial practicality of growing and harvesting microalgae for advanced biofuels production has not been demonstrated and faces many challenges including provisioning photobioreactors or open raceway ponds with large quantities of sunlight and CO_2, continuously harvesting microalgae, and extracting lipids from the microalgae. Neither has the commercial practicality of deoxygenating carbohydrate for the production of hydrocarbons been demonstrated. Although the U.S. DOE is committing significant resources to perfect both "algaculture" and "deoxygenation" of carbohydrates, our long history of growing carbohydrate crops and refining petroleum gives carbohydrate-derived hydrocarbon fuels a decided advantage in the near-term.

Sources of carbohydrate biomass include starch crops like maize, sugar crops like sugar cane, and cellulosic crops like switchgrass, corn stover, and wood residues. The recent controversy over diversion of corn from (primarily) animal feed to ethanol production has, in many people's minds, disqualified starch crops or other "food crops" from consideration as a future source of carbohydrate for biofuels production. The reality is that as long as the supply of corn exceeds the demand from food markets, it will be sold into "non-food" markets whether for the production of biofuels, bioplastics, or commodity chemicals.

Although sugar cane is also a food crop, its use as carbohydrate feedstock for biofuels production faces less resistance than corn. There are several reasons why sugar cane is viewed more favorably. First, it can be argued that sugar cane is not being grown on prime farmland at least compared to corn production, thus dodging the food vs. fuel criticism. Second, much of the expansion of sugar cane production has occurred on degraded farmland or abandoned pasture land, which mutes some of the early criticism that it is directly responsible for the destruction of Amazonian rainforests and increased greenhouse gas emissions (biofuels agriculture in the U.S. cannot as easily counter criticisms about its influence on land use change since the purported effects are all indirect and cannot be directly observed). The use of crop residue (bagasse) for process heat and electricity generation in sugar cane ethanol plants also results in an excellent lifecycle GHG score as calculated by the California Low Carbon Fuel Standard.[12] Finally, sugar cane biofuels have generally been cheaper to produce than starch-based biofuels, which helps deflect criticism of using a food crop for fuels production.

Ultimately, large-scale production of carbohydrate-derived biofuels will require exploitation of lignocellulosic biomass, whether collected as forestry or agricultural residues or purposely grown as dedicated herbaceous or woody energy crops. Among the lignocellulosic resources, forestry residues appears to raise the least objections, possibly because of the widespread incidence of dying forests due to pest infestation and a growing appreciation of the folly of letting deadwood accumulate in forests as wildfire hazards. Use of agricultural residues resolves the debate of food vs. fuel but raises other issues of sustainability as a result of increased potential for soil erosion, loss of soil carbon, and removal of nutrients with the removed residues.[13] Perennial plants, whether grasses or trees, grown as dedicated energy crops especially in polyculture address concerns about the environmental sustainability of biofuels agriculture but raises concerns about devoting agricultural land to non-food purposes (this objection overlooks the role agriculture has always played in non-food sectors of our economy). Nevertheless, as long as biofuels agriculture reduces the amount of carbon and nutrients in the soil compared to natural ecosystems, biofuels will be challenged for its environmental sustainability. Meeting this sustainability challenge will require new systems for biomass production and biofuels manufacture.

How will these energy sources be converted into energy carriers?

The future of biofuels is likely to be hydrocarbons very similar to petroleum-derived gasoline, diesel fuel, and aviation fuels except that it will be produced from lipids or carbohydrates. Despite this down selection on suitable energy carriers, an intriguing number of pathways to advanced biofuels remain. The lipid-based and carbohydrate-based pathways have uniquely different barriers to their commercial development. Conversion of lipids to hydrocarbons is relatively straightforward if sufficient biomass resources can be developed. On the other hand, the supply of carbohydrate in the form of lignocellulose is relatively certain if only processes that cost-effectively remove oxygen from carbohydrate can be developed.

Lipids are essentially hydrotreated in the same manner as petroleum is hydrotreated today. Although different catalysts are employed, the liquid feedstock is contacted with hydrogen at very high pressures, which breaks double carbon-carbon bonds and removes oxygen from the feedstock molecules. The resulting hydrocarbons are refined to get the desired mixtures of hydrocarbons for synthetic gasoline, diesel fuel, and aviation fuel. The process is commercially established and only requires sufficient supplies of lipid feedstock.

A wider array of pathways is open to carbohydrate feedstocks. Sugar cane provides both sugar and a lignocellulosic residue known as bagasse. Companies like LS9 and Amyris have genetically modified bacteria to ferment sugar to terpenes and other highly reduced compounds. Like lipids, these can be hydrotreated to hydrocarbons. Sugars can also be catalytically converted to hydrocarbons. One example is aqueous phase processing, under commercial development by Virent, which reacts solutions of sugar or sugar-derived compounds in the presence of catalysts to directly produce hydrocarbons. Lignocellulosic biomass, including the bagasse from sugar cane processing, can be converted to hydrocarbons by either biochemical or thermochemical processes. The biological process employs acids or enzymes to hydrolyze complex carbohydrates in plant fibers into glucose and other simple sugars. Like the disaccharides from sugar cane, these could be transformed into hydrocarbons by advanced fermentations currently under development.

Whereas simple sugars are the basis for biochemically generating biofuels from carbohydrate feedstocks, the thermochemical route allows for other "processing intermediates" in addition to sugars. The process of thermal gasification yields syngas, a mixture of carbon monoxide (CO) and hydrogen (H_2) that can be catalytically upgraded to hydrocarbons via the Fischer-Tropsch process. Fast pyrolysis, another thermochemical route, produces a mixture of highly oxygenated organic compounds known as bio-oil that can be catalytically upgraded to hydrocarbon fuels. Interestingly, under certain process conditions simple sugars can also be produced during fast pyrolysis, which could be catalytically or bio-catalytically converted to hydrocarbons.

Among the thermochemical pathways, fast pyrolysis of biomass followed by hydroprocessing the resulting bio-oil into renewable diesel and gasoline has two attributes that could speed its commercialization compared to gasification-based biofuels. First, bio-oil could be produced in distributed processing facilities, yielding bio-oil that is more readily shipped to a centralized upgrading facility than biomass.[14] Second, the bio-oil lends itself to upgrading in

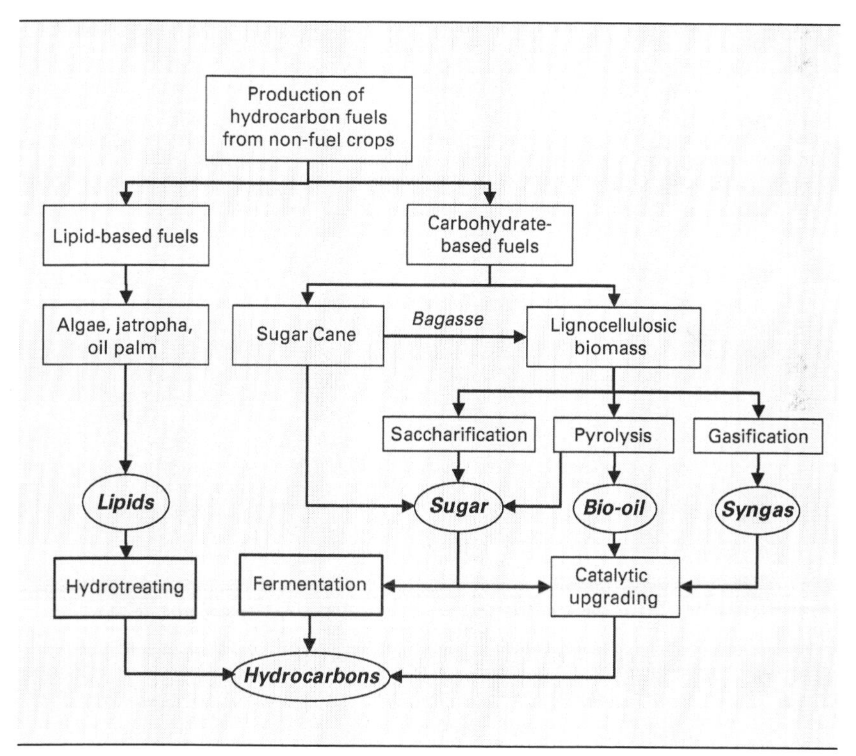

Pathways from "non-food" biomass feedstocks to hydrocarbon fuels.
Source: Author.

existing petroleum refineries,[15] which would avoid much of the high capital investment that currently hinders entry into advanced biofuels.

How quickly will "drop-in" biofuels be produced commercially from lipids or carbohydrates? In late 2006 ConocoPhillips began producing thousands of gallons per day of "green diesel" from soybean oil at a refinery in Ireland[16] although operations have since been discontinued until cheaper and more plentiful sources of lipids are identified. In early 2010, the airline industry began signing contracts for aviation fuels from both lipids and gasified cellulosic biomass.[17] This development is likely to encourage commercial production of gasoline and diesel fuel as well although the ultimate determinant will be the price of petroleum. Once petroleum prices rise decisively above $90 per barrel, wider deployment of advanced biofuels is likely to occur.

Can we devise a more rational approach to counting GHG emissions from biofuels?

The accounting of greenhouse gas emissions for renewable fuels, once thought to be well established, has been recently challenged by the concept of indirect land use change (ILUC). The Energy Independence and Security Act of 2007 (EISA) not only mandates the annual production of 36 billion gallons of renewable fuel by 2022, it requires these fuels to reduce greenhouse gas (GHG) emissions compared to petroleum-derived gasoline. A traditional accounting of greenhouse gas emissions directly attributable to the production and processing of corn to ethanol would meet the so-called "low carbon fuel standard" (LCFS). However EISA specifically requires GHG lifecycle analysis to consider "indirect emissions such as significant emissions from land use changes."[18] Indirect land use change raises the possibility that diverting agricultural land to support biofuels production could encourage conversion of natural ecosystems to agriculture in the developing world to replace this lost feed and food production. Conversion of land to agriculture is frequently accompanied by moderate to large carbon dioxide emissions from the mineralization of soil carbon in converted grasslands or the burning of rainforests.

The concept of ILUC falls short as a rational basis for regulating GHG emissions associated with agriculture. Field research demonstrates that GHG emissions associated with land-use change are driven by many cultural, technological, biophysical, political, economic, and demographic forces rather than by a single crop market.[19] It is essentially unknowable how people in the developing world will respond to the expanded use of biofuels in more developed countries, which makes problematic the assignment of GHG emissions associated with ILUC. Furthermore, ILUC provides no incentives to reduce GHG emissions since the biofuels industry is not able to influence land use outside the value-chain of its own feedstock suppliers while those directly responsible for converting natural ecosystems to agriculture are not penalized for doing so. One has to question the wisdom of adopting a policy that so grossly distorts responsibility for net GHG emissions that it is unlikely to be effective in reducing them. It seems short sighted not to expect both food and fuel agriculture to participate in efforts to moderate global climate change.

Fundamentally, the LCFS fails to address the fact that all economic activity generates GHG emissions. Under the proposed rules, only transportation fuels are held accountable for carbon discharged into the atmosphere. A potential solution to this dilemma is to develop and implement a system that acknowledges that carbon is "embedded" in goods and services.[20] Such an "Embedded Carbon Valuation System" (ECVS) accounts for the carbon accumulated along the value chain of a product's manufacture.[21]

Although no national inventory has been completed on the amount of GHG emissions associated with the production of various goods and services, they are not difficult to estimate on the basis of megagrams (metric tons) of carbon dioxide equivalence per $1000 of gross domestic product (Mg CO_2/$1000 GDP). For example, steel, concrete, and corn ethanol all produce about two tons of carbon dioxide

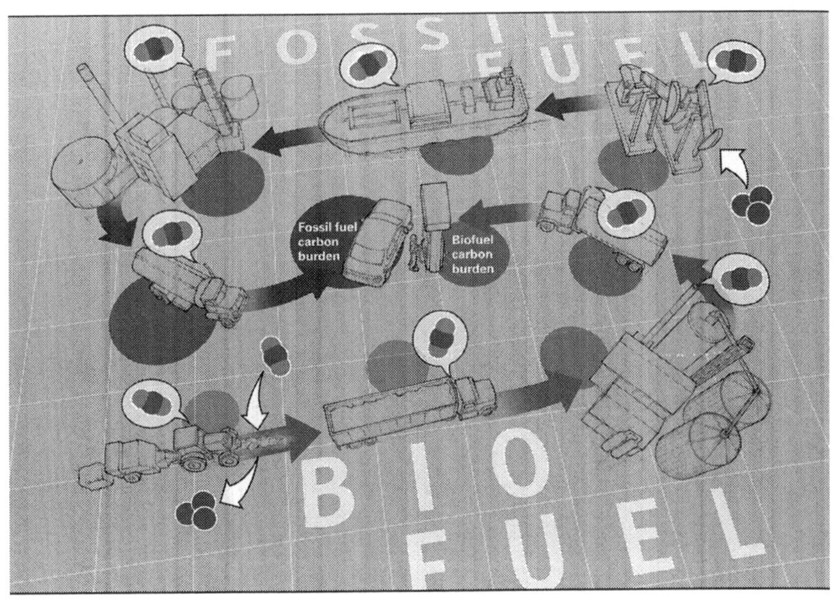

All economic activity generates greenhouse gas emissions. If all goods and services, not just fuels, were assigned embedded carbon valuations based upon cumulative greenhouse gas emissions along their value chains, companies would have both incentive and means to reduce greenhouse gas emissions and even sequester carbon. *Source: Author.*

emissions per $1000 GDP. Beef from corn-fed cattle is four tons, gasoline from petroleum is six tons, and electricity from coal is almost ten tons. Clearly, products and services other than transportation fuels play a role in GHG emissions, which the LCFS does not address.

When applied to the overall economy of a nation, the amount of GHG emitted per unit of GDP is referred to as "carbon intensity." In principle, it should include net emissions of greenhouse gases from all economic activity. In practice, data compiled by the U.S. Energy Information Agency (EIA) only accounts for emissions from fossil fuels.[22] For example, it does not account for CO_2 emissions associated with the calcination of limestone for production of cement or control of sulfur emissions at power plants;

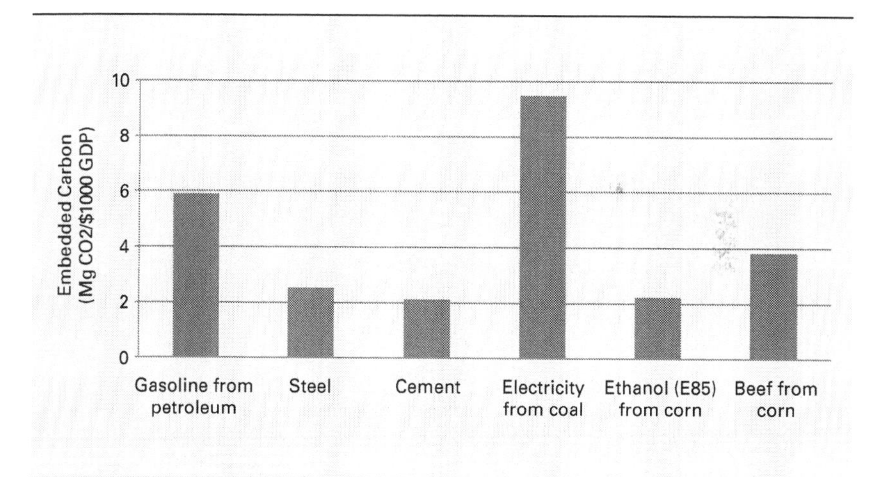

Greenhouse gas emissions for different kinds of economic activities can be compared on the basis of gross domestic product. The Low-Carbon Fuel Standard does not effectively address the ultimate sources of carbon being discharged into the atmosphere (all data based on direct land use change). *Source: Author.*

CO_2 and nitrous oxide (N_2O) emissions associated with certain kinds of agricultural practices; methane released from livestock production; or GHG emissions associated with products that are manufactured abroad and imported into a country. Nevertheless, it provides a snapshot on how effective a nation is in translating GHG emissions into economic activity. The carbon intensity for the U.S. economy, based on its indigenous greenhouse gas emissions from fossil fuel consumption, is only 0.5 Mg CO_2/\$1000 GDP. This reflects both the high service component of the U.S. economy as well as the fact that the U.S. outsources many of the most greenhouse gas intensive activities to other countries.[23] For example, China's economy, the origin of many U.S. imports, has a carbon intensity of 2.9 Mg CO_2/\$1000 GDP.

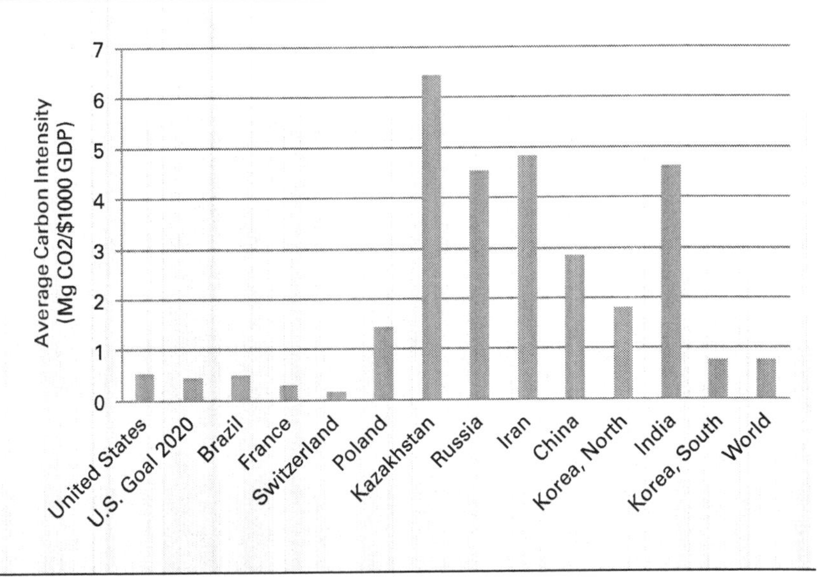

The carbon intensity of the U.S. economy is very low compared to most countries in the developing world. *Source: U.S. DOE EIA (2006).*

These differences in energy intensity among nations illustrates the fundamental inefficiency of assigning carbon emissions to the manufacturing facility rather than the manufactured product, which has the perverse effect of increasing manufacturing costs in countries that adopt greenhouse mitigation policies and encouraging U.S. companies to move their manufacturing base to countries with higher carbon intensities. Chinese and other foreign-based manufacturing would not have to pay this burden and would therefore have a competitive advantage. If we are to encourage economic growth while decreasing GHG emissions, we need to encourage the manufacture of products that have low embedded carbon valuations. Used in conjunction with a carbon tax, the ECVS would reduce industry's incentive to outsource by taxing all imports from non-carbon regulated economies at the same rate as domestically-produced products and services, based on the amount of emissions embedded in the product at the point of international transfer.

Another flaw of recently proposed carbon regulating systems (for example, California's LCFS, the American Clean Energy and Security Act of 2009, and the European Union's Emission Trading Scheme) is that they only regulate emissions generated by the transportation and utility sectors of the economy, which are together responsible for only one-third of U.S. GHG emissions.[24] Other sectors, including cement, food agriculture, and livestock production, are not restricted on their GHG emissions. In addition to greatly minimizing the impact of carbon regulation, the exclusion of these significant sources of GHGs would have a distortional impact on overall economic efficiency. By including all sources of GHG and placing them on a common basis (Mg CO2/$1000 GDP), the manufacture of products with higher economic value per unit of GHG emitted would be encouraged at the expense of products of lower value and higher GHG emissions. The ECVS could help solve the perplexing quandary of how to grow the economy while restricting GHG emissions.

Where should we grow all this biomass?

Stephen Chu, the U.S. Secretary of Energy, has formulated a vision for a "sugar economy" based on growing carbohydrate in the tropical parts of the world, converting it into sugars, and transporting it to energy markets around the world where it is converted into fuel.[25] More accurately, it should be called a "bioeconomy," which would admit lipids and lignin along with carbohydrate in the production of biofuels. U.S. agriculture likely views this global vision as a threat to their own economic interests while environmentalists may be alarmed at the prospect of corporate farming setting up business in the developing countries of the tropics. But it is an idea of tremendous merit.

The essential ingredients of a bioeconomy are sunlight, water, and carbon dioxide:

$$CO_2 + H_2O + \text{solar photons} \rightarrow CH_2O \text{ (biomass)} + O_2 \text{ (oxygen)}$$

Of these three ingredients, CO_2 is evenly distributed across the globe and the atmospheric supply seems secure for present. The supplies of sunlight and water, on the other hand, are not so evenly distributed. Not surprisingly, a solar irradiation map of the world reveals that with few exceptions, the sunniest parts of the world occur between the Tropic of Cancer and the Tropic of Capricorn.[26] Differences in cloud cover around the world distort this generalization with parts of the Amazon and Equatorial Africa receiving less sunlight than might otherwise be expected and parts of the U.S. Desert Southwest and the dry interior of Australia receiving more than expected. A rainfall map of the world shows the Eastern United States and Europe receiving plentiful rainfall along with virtually all of the tropics, where rainfall is often more than merely plentiful.[27] Merging the solar insolation and rainfall maps, it becomes clear that the tropics are the place to grow lots of biomass.[28]

In combination with sunlight and water a third factor makes the tropics a compelling location for biomass production: economic need. The United Nations has devised a measure of human welfare called the Human Development Index (HDI).[29] It is based on three indices: population health and longevity as measured by life expectancy at birth; knowledge and education as measured by adult literacy rates; and standard of living as measured by gross domestic

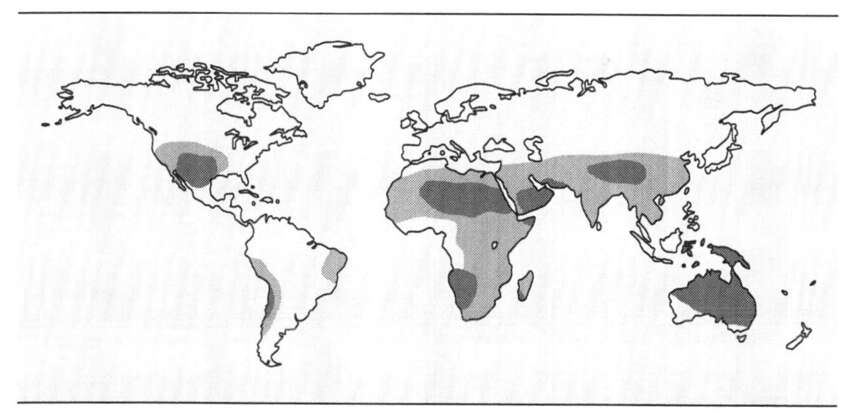

The sunniest places on Earth are concentrated in the tropics. Areas are shaded according to the annual average irradiance in watts per square meter (W/m^2). White: less than 200 W/m2; light gray: 200-250 W/m^2; dark gray: greater than 250 W/m^2. *Source: Redrawn from Reference 25.*

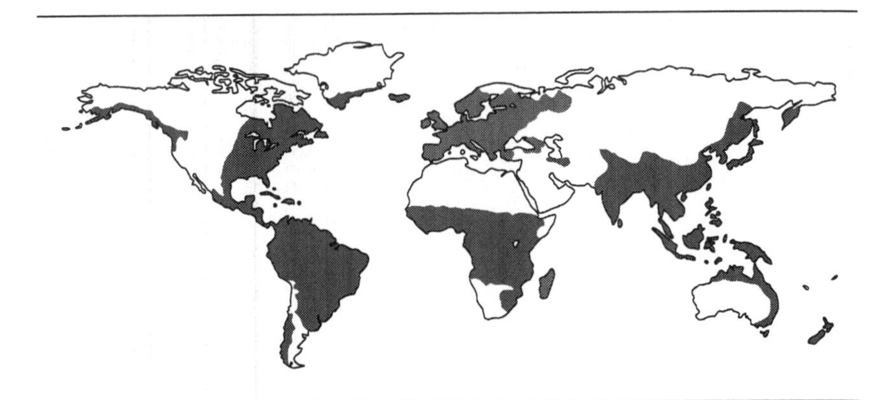

The rainiest places on Earth are concentrated in the tropics. Shaded areas receive at least 475 mm of annual rainfall with much of the tropics receiving over 1475 mm. *Source: Redrawn from Reference 26.*

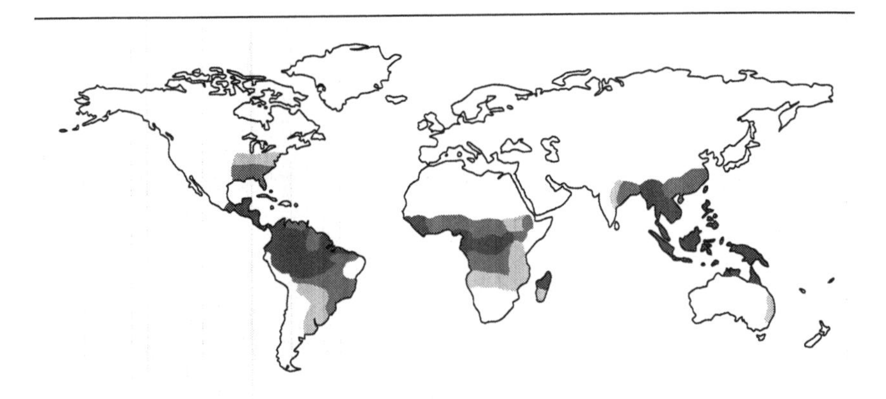

The convergence of sunlight and rainfall make the tropics the place to grow biomass. Areas are shaded according to net primary productivity in grams dry matter per square meter per year ($g/m^2/yr$). White: less than 1210 $g/m^2/yr$; light gray: 1210-1510 $g/m^2/yr$; medium gray: 1510-2010 $g/m^2/yr$; dark gray: 2010-2510 $g/m^2/yr$. *Source: Redrawn from Reference 27.*

product per capita at purchasing power parity. The lowest HDI's are concentrated in the tropics, especially in Africa and Southeast Asia. This will only improve with increasing economic opportunity for the people who live in these parts of the world.

In the tropics we find economic need, ecosystems critical to the health of the planet, and high potential for commercial biomass production. A system for producing biofuels that balances these three considerations would be a remarkable achievement. Unfortunately, tropical soils are among the most highly weathered and leached soils in the world. Most prominent among these are Oxisols ("oxidized" soils in the sense that they contain little soil carbon and excesses of iron and aluminum oxides) and Ultisols ("ultimate" soils,

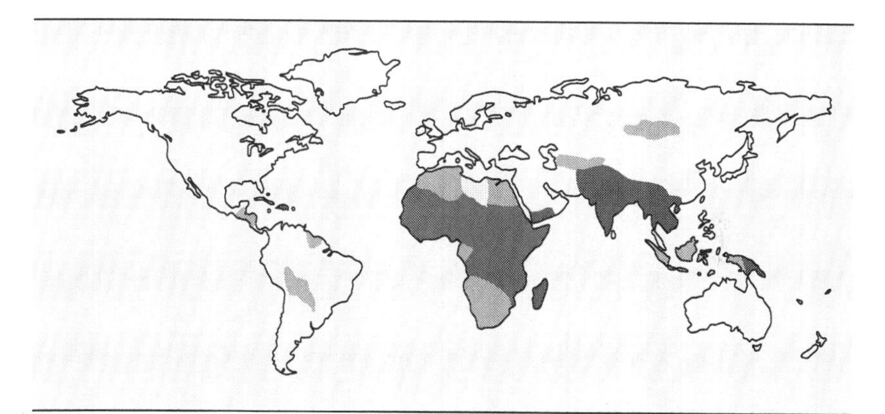

The most urgent need for advancing human welfare is concentrated in the tropics, especially Africa and Southeast Asia. The convergence of solar energy, precipitation, and human need in the tropics suggests an important role for biofuels in this part of the developing world. Areas are shaded according to the UN Human Development Index, which scales from 0 – 1.0. Dark gray: 0 – 0.5; light gray: 0.5 – 0.8; white: 0.8 – 1.0. *Source: Redrawn from Reference 28.*

in the sense of reaching the ultimate extent of weathering and leaching). Ironically, the combination of sunlight and water not only promotes the growth of biomass but exacerbates the loss of soil carbon (a process known as mineralization) and inorganic nutrients (leaching). The lush vegetation of the tropical zone is something of a veneer: remove it and what is left has poor prospects for sustained agriculture. In adapting to these soils, farmers historically practiced slash and burn agriculture, which replaces native vegetation with row crops for a few years until the supply of nutrients are depleted and the fields are abandoned for new lands. This practice might be acceptable for subsistence agriculture to support small populations, but it is both low yielding and wasteful of land resources and incompatible with sustainable biofuels agriculture.

A cure for the endemic problem of poor soils in the tropics may have been discovered in the tropics. Scattered through

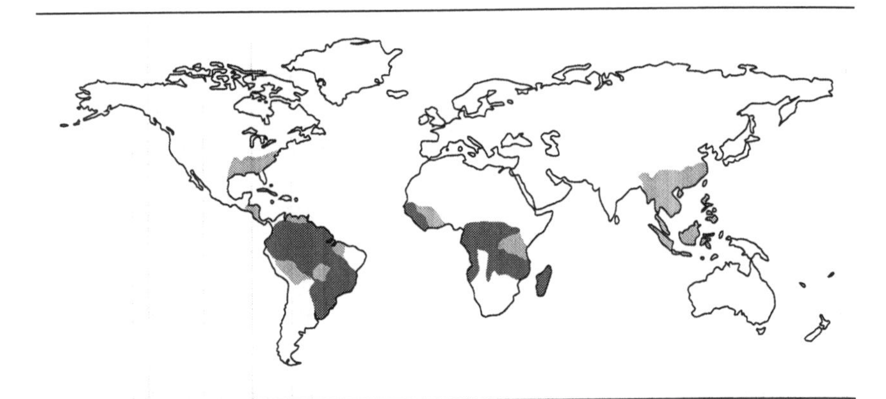

The tropics contain a disportionate share of the world's most highly weathered and leached soils such as Oxisols and Ultisols. Areas shaded according to soil type. Dark gray: Oxisols; light gray: Ultisols; white: all other soil types. Source: Reference 30.

the vast tracts of yellow and red Oxisols in the Amazon basin are patches of unusually dark earth (*Terra Preta* in Portuguese) that are also unusually fertile compared to the surrounding soil.[31] *Terra Preta* has been reported to be up to one hundred percent higher yielding than the surrounding natural soils. Indeed, local populations are reported to mine the patches of *Terra Preta*, which range in size from a few hectares to a few hundred hectares, for use as potting soil.

Terra Preta is characterized by high levels of soil carbon (as much as 10% compared to 0.5 – 1.0% for other good agricultural soils),[32] which is responsible for its distinctive dark color; high concentrations of plant nutrients including nitrogen, phosphorous, calcium, zinc, and manganese;[33] high cation exchange capacity,[34] which reduces nutrient leaching; and distinctive microbial activity,[35] which some believe contributes to its enhanced fertility.

Most intriguing of all, *Terra Preta* has anthropogenic (human-made) origins, created by pre-Columbian inhabitants of the Amazon basin who incorporated charred plant matter (charcoal), manure, shells, and fish and animal bones into the soil.[36] The intervention of humankind in its production is unmistakable: *Terra Preta* commonly contains large quantities of pottery and other artifacts. Although the creation of *Terra Preta* may have been unintentional, this seems unlikely as the process of charring biomass to form charcoal is more difficult than reducing biomass to inorganic ashes as in slash and burn soil management (some researchers who have studied *Terra Preta* refer to the soil making process as "slash and char" soil management). *Terra Preta* appears to have been generated between 450 BC and AD 950 at sites throughout the Amazon basin[37] as well as in other parts of the world. Thus these soils have persisted for millennia, offering an additional intriguing prospect of stably sequestering carbon in agricultural lands while enhancing the fertility of the soil.

The key to creating *Terra Preta* soils is the production and incorporation of large quantities of charcoal, as much as 25-50 tons per hectare, into soils.[38] Charcoal is generated through pyrolysis of biomass, which occurs at temperatures around 500°C and in the absence of oxygen. Charcoal kilns have been used to turn wood into charcoal for thousands of years, mostly for the purpose of manufacturing "smokeless" fuel for cooking and heating.[39] Traditional charcoal kilns have poor environmental reputations because of the smoke they emit and their role in the deforestation of Europe before the introduction of coal.[40] A modern system would capture or burn this smoke for more useful purposes. Slow pyrolysis, requiring several minutes or even many hours to complete, maximizes charcoal production and produces an energy-rich gas co-product called syngas or, more accurately, producer gas. Fast pyrolysis, which is accomplished in a few seconds, generates bio-oil as the main product with charcoal and gas representing co-products.

Charcoal produced for the specific intent of incorporation into agricultural lands is called biochar although its composition may be substantially the same as charcoal produced in traditional kilns. Unlike traditional kilns, though, the feedstock would not be cordwood from forests but crop residues or forest residues. Most often envisioned are relatively small pyrolysis plants distributed widely in agricultural landscapes to reduce the distance to transport biomass to the pyrolyzer and biochar back to the fields. Fortuitously, the biochar appears to contain a substantial fraction of the inorganic nutrients contained in the biomass, especially potassium and phosphorous, which suggests that biochar application will also recycle nutrients to farmland.[41]

This system of biochar agriculture and thermochemical biofuels production would allow developing countries in the tropics to participate in the emerging bioeconomy by addressing legitimate concerns about expanding agriculture into pristine rainforest environments. The best way to protect these natural ecosystems is to improve crop yields for tropical farmland and restore to productive use the estimated 12 – 25 million acres of farmland lost every year due to soil degradation.[42]

Why are we producing biofuels?

Ultimately, learning how to make advanced biofuels is our pathway to harnessing solar energy for the sustainable production of food and energy, both of which are essential to the economic life of the world's people. Those who argue that solar energy is not sufficiently efficient or economic should remember that the fossil energy that we exploit today is solar energy captured by photosynthesis eons ago and converted into petroleum, natural gas, and coal by geological processes. Undoubtedly we would declare fossil energy deposits as inefficient and uneconomic if nature had not done the hard work for us. As these stores are depleted, we will have to take responsibility for capturing and converting solar energy into replacements for fossil energy. We have little other choice.

Those who argue that agriculture should devote itself exclusively to food production overlook the fact that arable lands have always been used to provide a diversity of products to society including food, energy, chemicals, and materials. It was only in the past century that exploitation of fossil fuels made possible the shift from a bioeconomy that was essentially powered by solar energy to a petroleum economy that not only drove our increasing energy consumption but also became an integral part of food production. Clearly, harnessing solar energy for such multiple purposes will entail landscape-scale interventions in the biosphere that impact both human societies and natural ecosystems. As stewards of these interventions we should make sure that the inevitable impacts are positive for all inhabitants of the Earth's biosphere. That is the responsibility we assume.

Notes

Chapter 2

1. National Science and Technology Council (1997) Interagency Assessment of Oxygenated Fuels, Committee on Environment and Natural Resources, June.

2. Yacobucci, B. D. (2006) "Boutique Fuels" and Reformulated Gasoline: Harmonization of Fuel Standards, Congressional Research Service, Library of Congress. Available on the Web at: http://fpc.state.gov/documents/organization/67822.pdf (accessed December 31, 2008).

3. Blas, J., and Crooks, E. (2007) Drive on biofuels risks oil price surge, Financial Times, June 5. Available on the web at: http://www.ft.com/cms/s/0/aeb9a650-136e-11dc-9866-000b5df10621.html?nclick_check=1 (accessed April 9, 2009).

4. Lane, J. (2007) Venture capitalists invest $750 million in biofuels in 2007, Biofuels Digest, January 18. Available on the Web at: http://biofuelsdigest.com/blog2/2008/07/10/venture-capitalists-invest-280-million-in-advanced-biofuels-in-q2-08-84-million-for-algae-projects/trackback/ (accessed April 9, 2009).

5. Herndon, A. (2011) KiOR unchanged after pricing IPO at 29% below top of range, Bloomberg, June 24. Available on the web at: http://www.bloomberg.com/news/2011-06-24/kior-little-changed-after-pricing-ipo-29-below-top-of-range.html (accessed June 25, 2011).

6 Environmental Protection Agency (2009) EPA Proposes New Regulations for the National Renewable Fuel Standard Program for 2010 and Beyond, Regulatory Announcement. Available on the Web at: http://www.epa.gov/oms/renewablefuels/420f09023.pdf (accessed September 4, 2011).

7 Ferrett, G. (2007) Biofuels "crime against humanity," BBC News, October 27. Available on the web at: http://news.bbc.co.uk/2/hi/7065061.stm (accessed August 29, 2011).

8 Grocery Manufacturers Association (2008) GMA Statement Regarding Consumer Price Index Data for March 2008, April 16. Available on the Web at: http://www.gmabrands.com/news/docs/NewsRelease.cfm?DocID=1829 (accessed January 10, 2009).

9 Searchinger, T., Heimlich, R., Houghton, R. A., Dong, F., Elobeid, A., Fabiosa, J., Tokgoz, S., Hayes, D., and Yu, T.-H. (2008) Use of U.S. Croplands for Biofuels Increases Greenhouse Gases Through Emissions from Land-Use Change, Science 319 (5867) 1238 – 1240; originally published in Science Express, 7 February, DOI: 10.1126/science.1151861.

10 Fargione, J., Hill, J., Tilman, D., Polasky, S., and Hawthorne, P. (2008) Land Clearing and the Biofuel Carbon Debt, Science 319 (5867) 1235 – 1238; originally published in Science Express, 7 February, DOI: 10.1126/science.1152747.

11 H.R. 6 Energy Independence and Security Act of 2007. Available on the Web at http://www.govtrack.us/ congress/billtext.xpd?bill=h110-6 (accessed December 31, 2008).

12 Winter, A. (2010) New Biofuels Regs Could Still Face Fight From Capitol Hill, New York Times, February 4. Available on the Web at: http://www.nytimes.com/gwire/2010/02/04/04greenwire-new-biofuels-regs-could-still-face-fight-from-80948.html (accessed September 4, 2011).

13 Bevill, K. (2010) CARB to reduce ILUC for ethanol, Ethanol Producer Magazine, November 15. Available on the Web at: http://www.ethanolproducer.com/articles/7160/carb-to-reduce-iluc-for-corn-ethanol (accessed September 4, 2011).

14 Parker, M. and Whitten, D. (2010) U.S. EPA Slashes Cellulosic Ethanol Target for 2010, Bloomberg, February 3. Available on the web at: http://www.bloomberg.com/apps/news?pid=newsarchive &sid=asmNq69P7yGE (accessed August 29, 2011).

15 Wald, M. and Barrionuevo, A. (2007) A Renewed Push for Ethanol, Without the Corn, New York Times, April 17. Available on the web at: http://www.nytimes.com/2007/04/17/business/17ethanol.html (accessed August 29, 2011).

16 Klein-Marcuschamer, D., Oleskowicz-Popiel, P., Simmons, B.A., and Blanch, H.W. (2010) Technoeconomic analysis of biofuels: A wiki-based platform for lignocellulosic biorefineries, Biomass and Bioenergy 34 (12) 1914-1921. Available on the web at: http://www.lbl.gov/tt/publications/ 2678pub.pdf (accessed August 29, 2011).

17 Kazi, F.K., Fortman, J.A., Anex, R.P., Hsu, D.D., Aden, A., Dutta, A., and Kothandaraman, G. (2010) Techno-economic comparison of process technologies for biochemical ethanol production from corn stover, Fuel 89(Supplement 1) S20-S28.

18 Tao, L., et al. (2011) Process and technoeconomic analysis of leading pretreatment technologies for lignocellulosic ethanol production using switchgrass, Bioresource Technology DOI: 10.1016/j.biortech.2011.07.051.

Chapter 3

1 Energy Information Agency (2010) International Energy Outlook 2010. Available on the Web at: http://www.eia.gov/oiaf/ieo/index.html (accessed September 4, 2011).

2 Hubbert, M. K. (1956) Nuclear Energy and the Fossil Fuels, American Petroleum Institute Drilling and Production Practice, Proceedings of Spring Meeting, San Antonio.

3 Anon (2009) Petroleum Navigator, Energy Information Agency, Department of Energy, April 30. Available on the web at: http://tonto.eia.doe.gov/dnav/pet/hist/mttupus1A.htm (accessed May 14, 2009).

4 Minahan, T. (2007) Will Power Hungry China Boost Your Supply Bill? Supply Excellence, March 28. Available on the Web at: http://www.supplyexcellence.com/blog/2007/03/28/will-power-hungry-china-boost-your-supply-bill/ (accessed January 13, 2009).

5 Johnson, H. R., et al. (2004) Assessment of Strategic Issues, The Strategic Significance of America's Shale Oil Resource Volume I, Department of Energy, 2004. Available on the web at: http://www.unconventionalfuels.org/publications/reports/npr_strategic_significancev1.pdf (accessed May 9, 2009).

6 Worldwatch Institute (2005) State of the World 2005: Redefining Global Security, New York, p. 107.

7 Deffeyes, K. (2001) Hubbert's Peak: The Impending World Oil Shortage, Princeton University Press, Princeton, p. 8.

8 Smil, V. (2003) Energy at the Crossroads: Global Perspectives and Uncertainties, The MIT Press, Cambridge, p. 192.

9 Glover, P. (2008) Aramco Chief Debunks Peak Oil, Energy Tribune, January 17. Available on the web at: http://www.energytribune.com/articles.cfm?aid=764 (accessed May 9, 2009).

10 Foucher, S. (2006) Peak Oil Update – November 2006: Production Forecasts and EIA Oil Production Numbers, The Oil Drum, November 20. Available on the web at: http://www.theoildrum.com/ story/2006/11/13/225447/79 (accessed June 5, 2009).

11 BP (2010) Statistical Review of World Energy. Available on the Web at: http://www.bp.com/sectionbodycopy.do?categoryId=7500&contentId=7068481 (accessed September 4, 2011).

12 International Energy Agency (2010) World Energy Outlook 2010, November. Available on the Web at: http://www.worldenergyoutlook.org/docs/weo2010/WEO2010_ES_English.pdf (accessed September 4, 2011).

13 Morehouse, D. F. (1997) The Intricate Puzzle of Oil and Gas Reserves Growth, Natural Gas Monthly, July. Available on the Web at: http://tonto.eia.doe.gov/ftproot/features/morhouse.pdf (accessed January 13, 2009).

14 Anon (2009) Iran Plans to Produce Biofuels, Press TV, April 8. Available on the Web at: http://www.presstv.ir/detail.aspx?id=90788§ionid=351020103 (accessed April 10, 2009).

15 CIA (2008) Rank Order – Oil – Proved Reserves. The World Factbook, December 18. Available on the Web at: https://www.cia.gov/library/publications/the-world-factbook/rankorder/2178rank.html (accessed January 13, 2009).

16 Keeling, R. F., et al. (2008) Atmospheric CO2 values (ppmv) Derived from In situ Air Samples Collected at Mauna Loa, Hawaii, Scripps Institute of Oceanography, La Jolla, May. Available on the web at: http://cdiac.ornl.gov/ftp/trends/co2/maunaloa.co2 (accessed May 15, 2009).

17 Royal Society (2001) Joint Science Academies' Statement: The Science of Climate Change. Available on the Web at: http://royalsociety.org/displaypagedoc.asp?id=13619 (accessed February 3, 2009).

18 Solomon, S., Plattner, G.-K., Knutti, R., and Friedlingstein, P. (2009) Irreversible Climate Change Due to Carbon Dioxide Emissions, Proceedings of the National Academy of Sciences 106, 1704-1709. Available on the Web at: http://www.pnas.org/content/early/2009/01/28/0812721106.full. pdf+html (accessed February 4, 2009).

19 Broecker, W. S. and Kunzig, R. (2008) Fixing Climate, Hill and Wang, New York, p. 191.

20 Watson, R. T., et al. (2002) Climate Change 2001: Synthesis Report. Contribution to the Third Assessment Report of the Intergovernmental Panel on Climate Change, Cambridge University Press, Cambridge, United Kingdom.

21 Hansen, J., et al. (2007) Dangerous Human-Made Interference with Climate: A GISS Model Study, Atmos. Chem. Phys., 7, 2287–2312. Available on the Web at: www.atmos-chem-phys.net/7/2287/ 2007/ (accessed February 4, 2009).

22 Boden, T. A., Marland, G., and Andres, R. J. (2009) Global, Regional, and National Fossil-Fuel CO2 Emissions, Carbon Dioxide Information Analysis Center, Oak Ridge National Laboratory, U.S. Department of Energy, Oak Ridge, Tenn., U.S.A. doi 10.3334/CDIAC/00001. Available on the Web at: http://cdiac.ornl.gov/trends/emis/meth_reg.html (accessed August 26, 2011).

23 Reinhart, C.M. and Rogoff, K. (2009) This Time is Different: Eight Centuries of Financial Folly, Princeton University Press, Princeton.

24 Roubini, N. and Mihm, Stephen (2010) Crisis Economics: A Crash Course in the Future of Finance, Penguin, New York.

25 Dunaway, S. (2009) Global Imbalances and the Financial Crisis, Council Special Report No. 44, March. Available on the Web at: http://www.cfr.org/content/publications/attachments/Global_Imbalances_CSR44.pdf (accessed April 10, 2009).

26 Anon (2009) When a Flow Becomes a Flood, The Economist, January 24. Available on the Web at: http://www.economist.com/displaystory.cfm?story_id=12972083 (accessed April 8, 2009).

27 Rajan, R. (2010) Fault Lines: How Hidden Fractures Still Threaten the World Economy, Princeton University Press, Princeton.

28 Dudley, W.C. (2011) U.S. Economic Policy in a Global Context: Remarks by President Dudley at the Foreign Policy Association Corporate Dinnner, New York City. Federal Reserve Bank of New York, June 7. Available on the web at http://newyorkfed.org/newsevents/speeches/2011/ dud110607.html (accessed June 20, 2011).

29 Chan, S. (2010) U.S. Steps up Criticism of China's Practices, The New York Times, September 16, 2010. Available on the Web at: http://www.nytimes.com/2010/09/17/business/17geithner.html (accessed September 4, 2011).

30 Cooper, H. (2010) Obama Promises to Double Export Growth by 2015, The New York Times, January 28. Available on the Web at: http://www.nytimes.com/2010/01/29/business/29trade.html (accessed September 4, 2011).

31 Huber, P. and Mill, M. P. (2005) The Bottomless Well: The Twilight of Fuel, the Virtue of Waste, and Why We Will Never Run Out of Energy, Basic Books, New York.

Chapter 4

1 Ariza, L. M. (2000) Burning Times for Hot Fusion, Scientific American 282, 19-20.

2 EIA (2011) Crude Oil Proved Reserves, Reserves Changes, and Production, Energy Information Agency, February 10. Available on the Web at http://tonto.eia.doe.gov/dnav/pet/pet_crd_pres_a_epc0_r01_ mmbbl_a.htm (accessed October 16, 2011).

3 Tverberg, G. E. (2008) The Bakken Formation: How Much Will It Help?, The Oil Well, April 26. Available on the Web at: http://www.theoildrum.com/node/3868 (accessed October 16, 2011).

4 Anon (2004) Overview of Natural Gas, NaturalGas.Org. Available on the Web at: http://www.naturalgas.org/overview/overview.asp (accessed October 16, 2011).

5 Alsever, J. (2008) Pickens' Natural Gas Idea Picking Up Steam, MSNBC, October 21, 2008. Available on the Web at: http://www.msnbc.msn.com/id/27052462/ (accessed October 16, 2011).

6 Smead, R. G. and Pickering, G. B. (2008) North American Natural Gas Supply Assessment, American Clean Skies Foundation Report, prepared by Navigant Consulting, July 4. Available on the Web at: http://www.afdc.energy.gov/afdc/pdfs/ng_supply_assessment_2.pdf (accessed October 16, 2011).

7 Efstathiou Jr., J. and Chipman, K. (2011) Fracking: The Great Shale Gas Rush, Bloomberg Business Week, March 3 (Available on the Web at: http://www.businessweek.com/magazine/content/11_11/ b4219025777026.htm, accessed October 16, 2011).

8 Mastrangelo, E. (2007) An Analysis of Price Volatility in Natural Gas Markets, U.S. Department of Energy, Energy Information Administration, Office of Oil and Gas, August. Available on the Web at http://tonto.eia.doe.gov/ftproot/features/ngprivolatility.pdf (accessed April 9, 2009).

9 EIA (2011) Natural Gas Consumption by End Use, Energy Information Agency, September. Available on the Web at: http://205.254.135.24/dnav/ng/ng_cons_sum_dcu_nus_a.htm (accessed October 16, 2011).

10 EPA (2007) Greenhouse Gas Impacts of Expanded Renewable and Alternative Fuels Use, U.S. Environmental Protection Agency, Office of Transportation and Air Quality, EPA420-F-07-035, April. Available on the Web at http://www.epa.gov/otaq/renewablefuels/420f07035.pdf (accessed October 16, 2011).

11 Kosich, D. (2008) Repeal Sought for Ban on U.S. Govt. Use of CTL, Oil Shale, Tar Sands Generated Fuel, Mine Web, April 11. Available on the Web at: http://www.mineweb.com/mineweb/view/ mineweb/en/page38?oid=50551&sn=Detail (accessed October 16, 2011).

12 Tarka, T. J. (2009) Affordable, Low-Carbon Diesel Fuel from Domestic Coal and Biomass, National Energy Technology Center Report, DOE/NETL-2009/1349, January 14. Available on the Web at: http://www.netl.doe.gov/energy-analyses/pubs/CBTL%20Final%20Report.pdf (accessed October 16, 2011).

13 Johnson, K. (2008) Steven Chu: 'Coal is My Worst Nightmare,' Environmental Capital, The Wall Street Journal, December 11. Available on the Web at http://blogs.wsj.com/environmentalcapital/ 2008/12/11/steven-chu-coal-is-my-worst-nightmare/ (accessed October 16, 2011).

14 Ward, Jr., K. (2008) Energy Secretary Nominee Sees Coal as 'Nightmare,' Charleston Gazette, December 17. Available on the Web at http://www.windisgood.ca/site/chu_on_coal.pdf (accessed October 16, 2011).

15 Anon (1986) Hundreds Gassed in Cameroon Lake Disaster, BBC News, August 21. Available on the Web at: http://news.bbc.co.uk/onthisday/hi/dates/stories/august/21/newsid_3380000/3380803.stm (accessed October 16, 2011).

16 Anon (2004) Worldwide Look at Reserves and Production, Oil Gas J. 102, 22–23, December 18. Available on the Web at http://www.ogfj.com/display_article/115038/58/ARCHI/none/none/1/WORLDWIDE-LOOK-AT-RESERVES-AND-PRODUCTION/ (accessed April 9, 2009).

17 Charpentier, A. D., Bergerson, J. A., and MacLean, H. L. (2009) Understanding the Canadian Oil Sands Industry's Greenhouse Gas Emissions Environ. Res. Lett. 4, 1-11, January 20. Available on the Web at: http://www.iop.org/EJ/article/1748-9326/4/1/014005/erl9_1_014005.pdf (accessed October 16, 2011).

18 Bartis, J. T., LaTourette, L., Dixon, L., Peterson, D. J., and Cecchine, G. (2005) Oil Shale Development in the United States Prospects and Policy Issues, U.S. DOE Technical Report MG-414-NETL, prepared by the RAND Corporation.

19 Andrews, A. (2006) Oil Shale: History, Incentives, and Policy, Congressional Research Service, April 13. Available on the Web at: http://www.fas.org/sgp/crs/misc/RL33359.pdf (accessed October 16, 2011).

20 Tiikma, L., Johannes, I., and Pryadka, N. (2002) Co-Pyrolysis of Waste Plastics with Oil Shale, Proceedings. Symposium on Oil Shale 2002, Tallinn, Estonia: 76.

21 Brandt, A. R. (2007) Converting Green River Oil Shale to Liquid Fuels with the Alberta Taciak Processor: Energy Inputs and Greenhouse Gas Emissions, Energy and Resources Group, University of California, June 1. Available on the Web at: http://abrandt.berkeley.edu/shale/Brandt%20_ATP_ Energy_and_Emissions_dist.pdf (accessed October 16, 2011).

22 Brandt, A. R. (2008) Converting Oil Shale to Liquid Fuels: Energy Inputs and Greenhouse Gas Emissions of the Shell in Situ Conversion Process, Environ. Sci. Technol. 42 (19), 7489–7495.

23 Hoffmann, R. (2006), Old Gas, New Gas, 94, American Scientist, pp. 16–18

24 Kvenvolden, K. A. (1993) Gas Hydrates – Geological Perspective and Global Change, Reviews of Geophysics 31, 173-187.

25 Collett, T. (2001) Natural Gas Hydrates—Vast Resource, Uncertain Future, U.S. Geological Survey, USGS Fact Sheet FS–021–01, March. Available on the Web at: http://pubs.usgs.gov/fs/fs021-01/fs021-01.pdf (accessed October 16, 2011).

26 Milkov, A. V. (2004). "Global Estimates of Hydrate-Bound Gas in Marine Sediments: How Much is Really Out There?" Earth-Sci Rev 66 (3-4): 183–197.

27 Anon (2002) A Technology Roadmap for Generation IV Nuclear Energy Systems, U.S. Department of Energy Research Advisory Committee and the Generation IV International Forum, 03-GA50034, December.

28 Tester, J. W., Drake, E. M., Driscoll, M. J., Golay, M. W., and Peters, W. A., Sustainable Energy: Choosing Among Options, MIT Press, Cambridge, MA, pp. 361-406.

29 Penner, S. S., Seisera, R., and Schultz, K. R. (2008) Steps Toward Passively Safe, Proliferation-Resistant Nuclear Power, Progress in Energy and Combustion Science 34, 275–287.

30 Andrews, A. (2008) Nuclear Fuel Reprocessing: U.S. Policy Development, Congressional Research Service, Library of Congress, Order Code RS22542, March 27.

31 Lenzen, M. (2008) Life Cycle Energy and Greenhouse Gas Emissions of Nuclear Energy: A Review, Energy Conversion and Management 49 (2008) 2178–2199.

32 EIA (1993) Renewable Resources in the U.S. Electricity Supply, DOE/EIA-0561(92). Available on the Web at http://www.eia.doe.gov/cneaf/electricity/pub_summaries/renew_es.html (accessed October 16, 2011).

33 BP (20011) Statistical Review of World Energy 2011. Available on the Web at http://www.bp.com/ sectionbodycopy. do?category Id=7500&contentId=7068481 (accessed October 16, 2011).

34 Smil, V. (2003) Energy at the Crossroads, MIT Press, Cambridge, MA, pp. 239-296.

Chapter 5

1 Borman, G. L. and Ragland, K. W. (1998) Combustion Engineering, McGraw-Hill, Boston, pp. 30-46.

2 Griffiths, J. F. and Barnard, J. A. (1996) Flame and Combustion, CRC Press, Boca Raton.

3 U.S. Department of Energy (2000) Fuel Cell Handbook, Fifth Edition (CD-ROM). Morgantown, WV/Pittsburgh, PA., U. S. DOE National Energy Technology Laboratory.

4 Wright, M. and Brown, R. C. (2007) Comparative Economics of Biorefineries Based on the Biochemical and Thermochemical Platforms, Biofuels, Bioprocessing, and Biorefineries 1, 49-56.

5 Gifford, J. and Brown, R. (2011) Four economies of sustainable automotive transportation, Biofuels, Bioproducts and Biorefining 5, 293-304.

6 Myers, D. B., Ariff, G. D., James, B. D., and Kuhn, R. C. (2003) Hydrogen from Renewable Energy Sources: Pathway to 10 Quads for Transportation Uses in 2030 to 2050, Final Report to the U.S. Department of Energy, Hydrogen Program Office, Washington, D.C. Available on the Web at: http://www.eere.energy.gov/hydrogenandfuelcells/pdfs/iia11_myers.pdf (accessed October 23, 2011).

7 Romm, J. J. (2004) The Hype About Hydrogen: Fact and Fiction in the Race to Save the Climate, Island Press, Washington, D.C.

8 Alsever, J. (2008) Pickens' Natural Gas Idea Picking Up Steam, MSNBC, October 21, 2008. Available on the Web at: http://www.msnbc.msn.com/id/27052462/ (accessed October 23, 2011).

9 U.S. Energy Information Agency (2007) About U.S. Natural Gas Pipelines - Transporting Natural Gas, http://205.254.135.24/pub/oil_gas/natural_gas/analysis_publications/ngpipeline/fullversion.pdf (accessed October 23, 2011).

10 MacKenzie, J. and Avery, W. (1996) Ammonia fuel: The Key to Hydrogen-Based Transportation, Proceedings of the Energy Conversion Engineering Conference 3, 1761 – 1766.

11 Starkman, E., Newhall, H., Sutton, R. and Maguire, T. (1966) Ammonia as a Spark Ignition Engine Fuel: Theory and Application. SAE Paper No. 660155.

12 Sorenson, S. (2001) Dimethyl Ether in Diesel Engines: Progress and Perspectives, Journal of Engineering for Gas Turbines and Power 123, 652.

13 Gray Jr, C. and Alson, J. (1989) The Case for Methanol, Scientific American 261, 108-115.

14 Bailey, B. K. (1996) Performance of Ethanol as a Transportation Fuel, in Handbook on Bioethanol: Production and Utilization, Wyman, C. E. (Ed.), Taylor & Francis, Washington, DC.

15 Lynd, L. R., Cushman, J. H., Nichols, R. J., and Wyman, C. E. (1991) Fuel Ethanol from Cellulosic Biomass, Science 251, 1318-1323.

16 Klass, D. L. (1998) Biomass for Renewable Energy, Fuels, and Chemicals, Academic Press, San Diego.

17 Schoutens, G. H., Groot, W. J. and Hoebeek, J. B. W., (1986) Application of isopropanol-butanol-ethanol mixtures as an engine fuel, Proc Biochem 21, 30.

18 Brown, R. C. (2003) Biorenewable Resources: Engineering New Products from Agriculture, Blackwell Publishing, Ames, IA, pp. 45-48.

19 Krawczyk, T. (1999) Specialty Oils, INFORM - International News on Fats, Oils, and Related Materials, 10, 552.

20 Scrosati, B., and Garche, J. (2010) Lithium batteries: Status, prospects and future, Journal of Power Sources 195, 2419-2430.

21 BCG (2010) Batteries for Electric Car—Challenges, Opportunities, and the Outlook to 2020, The Boston Consulting Group. Available on the Web at: http://www.bcg.com/documents/file36615.pdf (accessed October 23, 2011).

22 Anon (2009) Electric Cars--How Much Does It Cost per Charge? Scientific American, March 13. Available on the Web at: http://www.scientificamerican.com/article.cfm?id=electric-cars-cost-per-charge (accessed October 23, 2011).

23 Anon (2010) Tesla Roadster, Wikipedia. Available on the Web at: http://en.wikipedia.org/wiki/ Tesla_Roadster (accessed October 23, 2011).

24 Yvkoff, L. (2009) 'BusinessWeek' ponders lithium ion market war, February 18, CNET News. Available on the Web at: http://news.cnet.com/8301-11128_3-10166184-54.html (accessed October 23, 2011)

25 Morris, D. (2003) A Better Way to Get from Here to There, Institute for Local Self-Reliance, December. Available on the Web at: http://www.newrules.org/electricity/betterway.pdf (accessed October 23, 2011).

26 Anon (2010) BYD Auto to sale F3DM electric to personal buyers, People's Daily Online, March 23. Available on the Web at: http://english.people.com.cn/90001/90778/90860/6927936.html (accessed October 23, 2011).

27 Chambers, Nick (2010) First Chevy Volt Reaches Customers, Will Out-Deliver Nissan in December, PluginCars.com, December 16. Available on the Web at: http://www.plugincars.com/first-chevy-volts-reach-customers-will-out-deliver-nissan-december-106575.html (accessed October 23, 2011).

28 Rauch, J. (2008) Electro-Shock Therapy, The Atlantic, July/August. Available on the Web at: http://www.theatlantic.com/doc/200807/general-motors (accessed October 23, 2011).

29 Anon (2007) Renewable Energy Technology Roadmap Up to 2020, Creating Markets for Renewable Energy Technologies, EU RES Technology Marketing Campaign. Available on the Web at: http://www.erec.org/fileadmin/erec_docs/Documents/Publications/Renewable_Energy_Technology_Roadmap.pdf (accessed October 30, 2011).

30 Anon (2001) Comparing the Benefits and Impacts of Hybrid Electric Vehicle Options, Electric Power Research Institute, July. Available on the Web at: http://ourenergypolicy.org/docs/9/Comparing_ Hybrid_Electric_Vehicle_Options.pdf (accessed October 23, 2011).

31 National Research Council (2010) Transitions to Alternative Transportation Technologies -- Plug-in Hybrid Electric Vehicles, The National Academy Press, Washington, D.C.

32 Anon (2010) Low voltage, The Washington Post, August 1. Available on the Web at: http://www.washingtonpost.com/wp-dyn/content/article/2010/07/31/AR2010073102589.html (accessed October 23, 2011).

Chapter 6

1 Goettemoeller, J. and Goettemoeller, A. (2007) Sustainable Ethanol: Biofuels, Biorefineries, Cellulosic Biomass, Flex-Fuel Vehicles, and Sustainable Farming for Energy Independence, Prairie Oak Publishing, Maryville, MO

2 Kovarik, W. (2008) Ethanol's First Century, XVI International Symposium on Alcohol Fuels. Available on the Web at: http://www.radford.edu/wkovarik/papers/International.History.Ethanol.Fuel.html (accessed October 25, 2011).

3 Rosillo-Calle, F. and Cortez, L. A. B. (1998) Towards PROALCOOL II – A Review of the Brazilian Bioethanol Programme, Biomass and Bioenergy 14, 115-124.

4 Anon (2011) Brazil Energy, Index Mundi. Available on the Web at: http://www.indexmundi.com/ energy.aspx?country=br&product=gasoline&graph=production+consumption (accessed October 25, 2011).

5 Riveras, I. (2011) Brazil ethanol exports to stay weak in 2011 – trade, Reuters, September 15. Available on the Web at: http://af.reuters.com/article/energyOilNews/idAFN1518155820100915? sp=true (accessed October 25, 2011).

6 Collit, R. and Ewing, R. (2011) Senators doubt WTO would uphold U.S. ethanol tariff, Reuters, January 10. Available on the Web at: http://www.reuters.com/article/2011/01/10/us-ethanol-brazil-mccain-idUSTRE7095P420110110 (accessed October 25, 2011).

7 MacDonald, T. (2004) Ethanol Fuel Incentives Applied in the U.S. – Reviewed from California's Perspective, International Oil, Gas & Energy Dispute Management, January.

8 O'Brien, D. and Woolverton, M. (2009) U.S. Ethanol Production, Imports and Stocks, Kansas State University Research and Extension. Available on the Web at: http://www.agmanager.info/energy/US_Ethanol_01-09.pdf (accessed October 25, 2011).

9 Beattie, A. (2009) Obama Says No Quick End to Ethanol Dispute, Financial Times, March 15. Available on the Web at: http://www.ft.com/intl/cms/s/0/7f40b194-1109-11de-994a-0000779fd2ac.html (accessed October 25, 2011).

10 Anon (2011) Dealing with the real, The Economist, August 6. Available on the Web at: http://www.economist.com/node/21525439 (accessed October 25, 2011).

11 Crooks, E. and Meyer, G. (2011) Brazilian imports of US ethanol soar, Financial Times, May 5. Available on the Web at: http://www.ft.com/intl/cms/s/0/f1486874-775d-11e0-824c-00144feabdc0.html#axzz1bpDdESYK (accessed October 25, 2011).

12 Krauss, C. (2011) Ethanol Subsidies Besieged, The New York Times, July 7. Available on the Web at: http://www.nytimes.com/2011/07/08/business/energy-environment/corn-ethanol-subsidies-may-be-in-jeopardy.html?_r=1&ref=ethanol (accessed July 25, 2011).

13 Watson, S. A. and Ramstad, P. E., eds. (1987) Corn: Chemistry and Technology, American Association of Cereal Chemists, St. Paul, MN.

14 Bothast, R. J. and Schlicher, M. A. (2005) Biotechnological processes for conversion of corn into ethanol, Applied Microbiology and Biotechnology 67, 19-25.

15 Liska, A.J., Yang, H.S., Bremer, V.R., Klopfenstein, T.J., Walters, D.T., Erickson, G.E., and Cassman, K.G. (2009) Improvements in Life Cycle Energy Efficiency and Greenhouse Gas Emissions of Corn-Ethanol, Journal of Industrial Ecology 13, 58-74.

16 Kwiatkowski, J., McAloon, A., Taylor, F., and Johnston, D. (2006) Modeling the Process and Costs of Fuel Ethanol Production by the Corn Dry-Grind Process, Industrial Crops & Products 23, 288-296.

17 Brown, R. C. (2003) Biorenewable Resources: Engineering New Products from Agriculture, Blackwell Publishing, Ames, IA, pp. 162-165.

18 McAloon, A., Taylor, F., Yee, W., Ibsen, K., and Wooley, R. (2000) Determining the Cost of Producing Ethanol from Corn Starch and Lignocellulosic Feedstocks. National Renewable Energy Laboratory, October. Available on the Web at: http://www.agmrc.org/media/cms/16_5299EA3DD888C.pdf (accessed October 25, 2011).

19 Wright, M. and Brown, R. C. (2007) Comparative Economics of Biorefineries Based on the Biochemical and Thermochemical Platforms, Biofuels, Bioprocessing, and Biorefineries 1, 49-56.

20 Blanchard, P. H. (1992) Technology of Corn Wet Milling and Associated Processes, Elsevier, NY.

21 Goldemberg, J. (2008) The Brazilian Biofuels Industry, Biotechnology for Biofuels 1, 2.

22 IEA (2007) Biofuel Production, IEA Energy Technology Essentials, January 2. Available on the Web at: http://www.iea.org/techno/essentials2.pdf (accessed October 25, 2011).

23 Knothe, G. (2001) Historical Perspectives on Vegetable Oil-Based Diesel Fuels, Inform 12, 1103–1107.

24 Rudnick, L. R. (2006) Synthetics, Mineral Oils, and Bio-based Lubricants, CRC Press, Boca Raton.

25 Van Gerpen, J. H., et al. (2007) Biodiesel: An Alternative Fuel for Compression Ignition Engines, 2007 Agricultural Equipment Technology Conference. Available on the Web at: http://asae.frymulti.com/data/pdf/6/bcie2007/LecSeries31.pdf (accessed October 25, 2011).

26 Ouellette, R. (1998) Organic Chemistry: A Brief Introduction, Prentice Hall, New Jersey.

27 Lipinsky, E. S., McClure, T. A., Kresovich, S., Otis, J. L., and Wagner, C. K., (1984) Fuels and Chemicals from Oilseeds, in AAAS Selected Symposium, Westview Press, Boulder.

28 Van Gerpen, J. (2005) Biodiesel Processing and Production, Fuel Processing Technology 86, 1097-1107.

29 Brown, R. C. (2003) Biorenewable Resources: Engineering New Products from Agriculture, Blackwell Publishing, Ames, IA, pp. 179-182.

30 Krawczyk, T. (1999) Specialty Oils – An Unfilled Promise, International News on Fats, Oils, and Related Materials, 10, 552-561.

Chapter 7

1 Ferrett, G. (2007) Biofuels "crime against humanity," BBC News, October 27. Available on the Web at: http://news.bbc.co.uk/2/hi/7065061.stm (accessed October 25, 2011).

2 Grunwald, M. (2008) The Clean Energy Scam, Time Magazine, March 27. Available on the Web at: http://www.time.com/time/magazine/article/0,9171,1725975,00.html (accessed October 24, 2011).

3 Bailey, B. K. (1996) Performance of ethanol as a transportation fuel, in Handbook on bioethanol: production and utilization, Taylor and Francis, Bristol, Pennsylvania.

4 Lynd, L., Cushman, J. H., Nichols, R. J., and Wyman, C. E. (1991) Fuel Ethanol from Cellulosic Biomass, Science 251, 1318-1323.

5 Tyner, W. (2009) Big time issues facing the ethanol industry, Ethanol Today, October, 44-46. Available on the Web at: http://www.ethanoltoday.com/index.php?option=com_content&task=view&id=5&Itemid=6&fid=65 (accessed October 25, 2011).

6 Johnson, C. and Melendez, M. (2007) E85 Retail Business Case: When and Why to Sell E85, National Renewable Energy Laboratory, NREL/TP-540-41590, December. Available on the Web at: http://www.afdc.energy.gov/afdc/pdfs/41590.pdf (accessed October 25, 2011).

7 U.S. Environmental Protection Agency (2009) EPA Proposes New Regulations for the National Renewable Fuel Standard Program for 2010 and Beyond, EPA-420-F-09-023, May. Available on the Web at: http://www.epa.gov/OMS/renewablefuels/420f09023.htm#11 (accessed October 25, 2011).

8 Krawczyk, T. (1999) Specialty Oils – An Unfilled Promise, International News on Fats, Oils, and Related Materials, 10, 552-561.

9 Weyenberg, T. (2007 Improving Biodiesel Handling and Operability in Cold Weather, Biodiesel Magazine, October. Available on the Web at: http://www.biodieselmagazine.com/article.jsp?article_id=1866&q=&page=1 (accessed October 25, 2011).

10 Relerford, P. (2009) Stalled Minnesota school buses fuel biodiesel mandate debate: When low temps gelled fuel in school buses some wondered if biodiesel contributed to the problem, Minneapolis-St. Paul Star Tribune. Available on the Web at: http://www.startribune.com/local/ south/37748654.html (accessed October 25, 2011).

11 For example: Hill, J., et al. (2006) Environmental, Economic, and Energetic Costs and Benefits of Biodiesel and Ethanol Biofuels, June 2. Available on the Web at: http://www.pnas.org/content/ 103/30/11206.abstract (accessed October 25, 2011).

12 For example: Conniff, R. (2007) Who's Fueling Whom? Why the biofuels movement could run out of gas, Smithsonian magazine, November. Available on the Web at: http://www.smithsonianmag.com/ science-nature/presence-biofuel-200711.html (accessed October 25, 2011).

13 U.S. Department of Agriculture, Economic Research Service, Feed Grains Database: Yearbook Tables. Available on the Web at: http://www.ers.usda.gov/Data/FeedGrains/FeedYearbook.aspx#FSI (accessed October 25, 2011).

14 Schill, S. R. (2007) 300-Bushel Corn is Coming, Ethanol Producer Magazine, October. Available on the Web at: http://www.ethanolproducer.com/article.jsp?article_id=3330 (accessed October 25, 2011).

15 U.S. Department of Agriculture, National Agriculture Statistics Service, Data and Statistics. Available on the Web at: http://www.nass.usda.gov/QuickStats/index2.jsp#top (accessed October 25, 2011).

16 Hofstrand, D. (2008) Tracking the Profitability of Corn Production, Ag Decision Maker, Iowa State University Extension. Available on the Web at: http://www.extension.iastate.edu/agdm/crops/html/a1-85.html (accessed October 25, 2011).

17 Congressional Budget Office (1983) Agricultural export markets and the potential effects of export subsidies, Staff Working Paper, June. Available on the Web at: http://www.cbo.gov/doc.cfm?index=5024&type=0 (accessed October 25, 2011).

18 U.S. Department of Agriculture, National Agriculture Statistics Service Number of Farms, Land in Farms, and Average Farm Size, Iowa by County, 2006-2007. Available on the Web at: http://www. nass.usda.gov/Statistics_by_State/Iowa/Publications/Annual_Statistical_Bulletin/2008/4-5_08.pdf (accessed October 25, 2011).

19 Palmer, A. (2008) Beating Up on Ethanol: Glover Park Helps Frame the Debate, Roll Call, May 14. Available on the Web at: http://www.rollcall.com/issues/53_137/news/23620-1.html (accessed October 25, 2011).

20 Walt, V. (2008) The World's Growing Food-Price Crisis, Time Magazine, February 27. Available on the Web at: http://www.time.com/time/world/article/0,8599,1717572,00.html (accessed October 25, 2011).

21 Mitchell, D. (2008) A Note on Rising Food Prices, World Bank Policy Research Working Paper No. 4682, World Bank Development Prospects Group. Available on the Web at: http://papers.ssrn.com/ sol3/papers.cfm?abstract_id=1233058# (accessed October 25, 2011).

22 Monbiot, G. (2007) If We Want to Save the Planet, We Need a Five-Year Freeze on Biofuels, The Guardian, March 27. Available on the Web at: http://www.guardian.co.uk/commentisfree/2007/mar /27/comment.food (accessed October 25, 2011).

23 International Starch Institute, Maize (corn). Available on the web at: http://www.starch.dk/isi/starch/maize.asp (accessed October 25, 2011).

24 Hansen, R. and Huntrods, D. (2009) White Corn Profile, Agriculture Marketing Resource Center, Available on the Web at: http://www.agmrc.org/commodities__products/grains__oilseeds/corn_grain/white_corn_profile.cfm (accessed October 25, 2011).

25 United States Department of Agriculture (2009) Economics, Statistics and Market Information System, USDA Long-Term Agricultural Projection Tables, February. Available on the Web at: http://usda.mannlib.cornell.edu/MannUsda/viewStaticPage.do?url=http://usda.mannlib.cornell.edu/usda/ers/94005/./2009/index.html (accessed October 25, 2011).

26 U.S. Department of Agriculture (2009) National Agriculture Statistics Service, Livestock and Animals. Available on the Web at: http://www.nass.usda.gov/QuickStats/indexbysubject.jsp?Text1=&site= NASS_MAIN&select=Select+a+State&Pass_name=Cattle++All&Pass_group=Livestock+%26+Animals&Pass_subgroup=Poultry&list=Cattle++All#top (accessed October 25, 2011).

27 Author. These calculations are based on corn calories consumed in producing livestock or producing ethanol. Similar values are reported in Gillespie, J. R. (2004) Modern livestock & poultry production, Delmar Thomson Learning Cengage Learning, Clifton Park, NJ. The analysis does not include fossil fuel consumption by either industry, which is discussed in a later section.

28 U.S. Department of Agriculture (2008) Economic Research Service, Briefing Rooms, Food Marketing System in the U.S.: Price Spreads from Farm to Consumer. Available on the Web at: http://www.ers. usda.gov/Briefing/FoodMarketingSystem/pricespreads.htm (accessed October 25, 2011).

29 Ryan, M. (2008) White House sees food prices high for 2-3 years, Reuters, May 13. Available on the Web at: http://www.reuters.com/article/politicsNews/idUSWBT008982200805l3?feedType=RSS &feedName=politicsNews (accessed October 25, 2011).

30 Baffes, J. and Haniotis, T. (2010) Placing the 2006/08 Commodity Price Boom into Perspective, Policy Research Working Paper 5371, The World Bank, July. Available on the Web at: http://www-wds. worldbank.org/external/default/WDSContentServer/IW3P/IB/2010/07/21/000158349_20100721110120/Rendered/PDF/WPS5371.pdf (accessed October 25, 2011).

31 Lee, D. J. (2009) With corn prices in retreat, fuel vs. food debate loses heat, The Register-Guard, Jan 11. Available on the Web at: http://bioenergyuiuc.blogspot.com/2009/01/with-corn-prices-in-retreat-fuel-vs.html (accessed October 25, 2011).

32 Ranney, J. W. and Mann, L. K. (1994) Environmental considerations in energy crop production, Biomass and Bioenergy, 6, 211 - 228.

33 Hohenstein, W. G. and Wright, L. L. (1994) Biomass energy production in the United States: An overview, Biomass and Bioenergy, 6, 161 -173.

34 Lang, S. S. (2006) 'Slow, Insidious' Soil Erosion Threatens Human Health and Welfare as Well as the Environment, Cornell Study Asserts, March 20. Available on the Web at: http://www.news.cornell. edu/ stories/March06/soil.erosion.threat.ssl.html (accessed October 25, 2011).

35 Spiro, T. G. and Stigliani, W. M. (1996) Chemistry of the Environment, Prentice Hall, Upper Saddle River, NJ.

36 Steil, M. (2005) The ethanol equation, Minnesota Public Radio, March 21, available on the Web at http://news.minnesota.publicradio.org/features/2005/03/21_steilm_ethanolenergy/ (accessed October 25, 2011).

37 Estimated from data found in Shapouri, H., Duffield, J. A., and Wang, M. (2003) The Energy Balance of Corn Ethanol Revisited, Transactions of the American Society of Agricultural Engineers 46, 959-968.

38 Lovett, R. A. (2005) Turning corn into ethanol may not be worth it, San Diego Union-Tribune, August 3. Available on the Web at http://www.signonsandiego.com/uniontrib/20050803/news_lz1c03fuel. html (accessed October 25, 2011).

39 Pimentell, D. and Patzek, T. W. (2005) Ethanol Production Using Corn, Switchgrass, and Wood; Biodiesel Production Using Soybean and Sunflower, Natural Resources Research 14, 65-76.

40 Shapouri, H., Duffield, J. A., and Wang M. (2003) The Energy Balance of Corn Ethanol Revisited, Transactions of the American Society of Agricultural Engineers 46, 959-968.

41 Wang, M. (2005) Energy and Greenhouse Gas Emissions Impacts of Fuel Ethanol, National Corn Growers Renewable Fuels Forum, National Press Club, Washington DC, August 23. Available on the Web at: http://www.anl.gov/Media_Center/News/2005/news050823.html (accessed October 27, 2011).

42 Hunt, S. C., et al. (2006) Biofuels for Transportation, World Watch Institute, June 7. Available on the Web at: http://www.worldwatch.org/system/files/EBF038.pdf (accessed October 25, 2011).

43 Intergovernmental Panel on Climate Change (2000) Land Use, Land-Use Change, and Forestry; Special Report of the IPCC. Cambridge University Press, Cambridge.

44 Liska, A., Yang, H., Bremer, V., Klopfenstein, T., Walters, D., Erickson, G., and Cassman, K. (2009) Improvements in Life Cycle Energy Efficiency and Greenhouse Gas Emissions of Corn-Ethanol, Journal of Industrial Ecology 13, 58-74.

45 Smeets, E., Junginger, M., Faaij, A., Walter, A., and Dolzan, P. (2006) Sustainability of Brazilian bio-ethanol, Copernicus Institute at Universiteit Utrecht and Universidade Estadual de Campinas, Report NWS-E-2006-110, ISBN 90-8672-012-9. Available on the Web at: http://igiturarchive.library.uu.nl/ chem/2007-0628-202408/UUindex.html (accessed October 25, 2011).

46 Searchinger, T., Heimlich, R., Houghton, R. A., Dong, F., Elobeid, A., Fabiosa, J., Tokgoz, S., Hayes, D., and Yu, T.-H. (2008) Use of U.S. Croplands for Biofuels Increases Greenhouse Gases Through Emissions from Land-Use Change, Science 319, 1238–1240; originally published in Science Express, 7 February, DOI: 10.1126/science.1151861.

47 Fargione, J., Hill, J., Tilman, D., Polasky, S., Hawthorne, P. (2008) Land Clearing and the Biofuel Carbon Debt, Science Express, 7 February, DOI: 10.1126/science.1152747.

48 California Air Resource Board (2009) Staff Report: Initial Statement of Reasons. Proposed Regulation to Implement the Low Carbon Fuel Standard, Volume I, Subchapter 10. Climate Change, Article 4. Regulations to Achieve Greenhouse Gas Emission Reductions, Subarticle 7. Low Carbon Fuel Standard, March 5. Available on the Web at: http://www.arb.ca.gov/regact/2009/lcfs09/ lcfsisor1.pdf (accessed October 25, 2011).

49 U.S. Environmental Protection Agency (2009) EPA Life Cycle Analysis of Greenhouse Gas Emissions from Renewable Fuels, Office of Transportation and Air Quality, EPA-420-F-09-024, May. Available on the Web at: http://www.epa.gov/otaq/renewablefuels/420f09024.htm (accessed October 25, 2011).

50 Dumortier, J., Hayes, D. J., Carriquiry, M., Dong, F., Du, X., Elobeid, A., Fabiosa, J. F., and Tokgoz, S. (2009) Sensitivity of Carbon Emission Estimates from Indirect Land-Use Change, Center for Agricultural and Rural Development, Working Paper 09-WP 493. Available on the Web at: http://www.card.iastate.edu/publications/dbs/pdffiles/09wp493.pdf (accessed October 25, 2011).

51 Kline, K. L. and Dale, V. H. (2008) Biofuels: Effects on land and fire; Letter to the editor, Science 321, 199.

52 Food and Agriculture Organization (2010) Global Forest Resources Assessment 2010. Available on the Web at: http://www.fao.org/docrep/013/i1757e/i1757e.pdf (accessed June 26, 2011).

53 Food and Agriculture Organization (2005) State of the World's Forests 2005. Available on the Web at: http://www.fao.org/docrep/007/y5574e/y5574e00.htm (accessed October 25, 2011).

54 Rosenthal, E. (2009) New Jungles Prompt a Debate on Rain Forests, New York Times, January 29. Available on the Web at: http://www.nytimes.com/2009/01/30/science/earth/30forest.html (accessed October 25, 2011).

55 EPA (2010) EPA finalizes regulations for the National Renewable Fuel Standard program for 2010 and beyond, Office of Transportation and Air Quality, EPA-420-F-10-007, February. Available on the Web at: http://www.epa.gov/otaq/renewablefuels/420f10007.pdf (accessed October 25, 2011).

56 EPA (2010) EPA life cycle analysis of greenhouse gas emissions from renewable fuels, Office of Transportation and Air Quality, EPA-420-F-10-006, February. Available on the Web at: http://www.epa.gov/otaq/renewablefuels/420f10006.pdf (accessed October 25, 2011).

Chapter 8

1 Klass, D. L. (1998) Biomass for Renewable Energy, Fuels, and Chemicals, Academic Press, San Diego, pp. 339-344.

2 Miles, D. (2008) Military Looks to Synthetics, Conservation to Cut Fuel Bills, American Forces Press Service, June 6. Available on the Web at: http://www.defenselink.mil/news/newsarticle.aspx?id= 50131 (accessed October 26, 2011).

3 Kram, J. W. (2009) Aviation alternatives, Biodiesel Magazine, January. Available on the Web at: http://www.biodieselmagazine.com/article.jsp?article_id=3071 (accessed October 26, 2011).

4 USDA (2005) Agricultural Statistics 2004, Table 3-51. Available on the Web at: http://www.nass.usda.gov/Publications/Ag_Statistics/2004/index.asp (accessed October 25, 2011).

5 Hughes, S. (2011) Girls Scouts to reduce use of palm oil in cookies after two Juniors start campaign, The Washington Post, September 30. Available on the Web at: http://www.washingtonpost.com/ blogs/blogpost/post/girl-scouts-to-reduce-use-of-palm-oil-in-cookies-after-two-juniors-start-campaign/2011/09/30/gIQAbbiXAL_blog.html (accessed October 25, 2011).

6 Greenpeace UK (2008) FAQ: Palm oil, forests and climate change. Available on the Web at: http://www.greenpeace.org.uk/forests/faq-palm-oil-forests-and-climate-change (accessed October 25, 2011).

7 Naylor, R., et al. (2007) The ripple effect: Biofuels, food security, and the environment, Environment 49, November. Available on the Web at: http://www.environmentmagazine.org/ Archives/Back%20Issues/November%202007/Naylor-Nov07-full.html (accessed October 26, 2011).

8 Alrios, M. G., Tejado, A., Blanco, M., Mondragon, I., Labidi, J. (2008) Agricultural palm oil tree residues as raw material for cellulose, lignin, and hemicelluloses production by ethylene glycol pulping process, Chemical Engineering Journal, July. DOI:10.1016/j.cej.2008.08.008.

9 Anon (2008) Palm oil, Wikipedia. Available on the Web at: http://en.wikipedia.org/wiki/Palm_oil (accessed October 26, 2011).

10 Pae, P. (2009) Continental airlines uses biofuel on test flight, Los Angeles Times, January 8. Available on the Web at: http://articles.latimes.com/2009/jan/08/business/fi-biofuel8 (accessed October 26, 2011).

11 Fitzgerald, M. (2008) India's big plans for biodiesel, Technology Review, December 27. Available on the Web at: http://www.technologyreview.com/Energy/17940/ (accessed October 26, 2011).

12 Barta, P. (2007) Jatropha plant gains steam in global race for biofuels, Wall Street Journal, August 24. Available on the Web at: http://online.wsj.com/article/SB118788662080906716.html (accessed October 26, 2011).

13 Lane, J. (2009) The blunder crop: A Biofuels Digest special report on jatropha biofuels development. Biofuels Digest, March 24. Available on the Web at: http://biofuelsdigest.com/blog2/2009/03/24/ the-blunder-crop-a-biofuels-digest-special-report-on-jatropha-biofuels-development/ (accessed October 26, 2011).

14 Koonin, S. (2006) Getting serious about biofuels, Science 311, 435.

15 Glenn, E. P., et al. (1998) Irrigating crops with seawater, Scientific American, August: 76-81. Available on the Web at: http://www.miracosta.edu/home/kmeldahl/articles/crops.pdf (accessed October 26, 2011).

16 Glenn, E. P., et al. (1997) Water requirements for cultivating Salicornia bigelovii Torr. with seawater on sand in a coastal desert environment, Journal of Arid Environments 36, 711-730.

17 Glenn, E. P., et al. (1991) Salicornia bigelovii Torr.: An Oilseed Halophyte for Seawater Irrigation, Science 251(4997): 1065-1067.

18 Christianson, R. C. (2008) Sea asparagus can be oilseed feedstock for biodiesel, Biomass Magazine, July 13. Available on the Web at: http://www.biomassmagazine.com/article.jsp?article_id=1864 (accessed October 26, 2011).

19 Dickerson, M. (2008) A voice in the desert, Seattle Times, July 25. Available on the Web at: http://seattletimes.nwsource.com/html/nationworld/2008071982_salicornia25.html (accessed October 26, 2011).

20 Mata, T.M., Martins, A.A., Caetano, N.S. (2010) Microalgae for biodiesel production and other applications: A review, Renewable and Sustainable Energy Reviews 14: 217-232.

21 Richmond, A. (2004) Handbook of microalgal culture: biotechnology and applied phycology. Blackwell Science Ltd.

22 Sheehan, J., Dunahay, T., Benemann, J., Roessler, P. (1998) A look back at the U.S. Department of Energy's Aquatic Species program, National Renewable Energy Laboratory, NREL/TP-580-24190, July. Available on the Web at: www.nrel.gov/biomass/pdfs/24190.pdf (accessed October 26, 2011).

23 Lane, J. (2008) Algae companies: Biofuels and biomass technologies database, Biofuels Digest, October 8. Available on the Web at: http://www.biofuelsdigest.com/blog2/2008/10/08/algae-companies-biofuels-and-biomass-technologies-database/ (accessed October 26, 2011).

24 Barron, R. (2008) Algae biofuel investments explode, Greentech Media, September 18. Available on the Web at: https://www.greentechmedia.com/articles/read/algae-biofuel-investments-explode-1434/ (accessed October 26, 2011).

25 Stein, M.L. (2009) The summer of algae, The Wall Street Journal, July 14. Available on the Web at: http://blogs.wsj.com/venturecapital/2009/07/14/the-summer-of-algae/ (accessed November 20, 2011).

26 Bevill, K. (2008) Aquaflow achieves success harvesting wild algae, Biomass Magazine, April 11. Available on the Web at: http://www.biomassmagazine.com/article.jsp?article_id=1562 (accessed October 26, 2011).

27 Kanellos, M. (2008) Inside Sapphire's algae-fuel plans, Greentech Media, October 13. Available on the Web at: http://greenlight.greentechmedia.com/2008/10/13/inside-sapphires-algae-fuel-plans-646/ (accessed October 26, 2011).

28 Kanellos, M. (2009) Algae biodiesel: It's $33 a gallon, Greentech Media, February 3. Available on the Web at: http://www.greentechmedia.com/articles/read/algae-biodiesel-its-33-a-gallon-5652/ (accessed November 20, 2011).

29 Associated Press (2008) Algae emerges as potential fuel source, New York Times, December 1. Available on the Web at: http://www.nytimes.com/2007/12/02/us/02algae.html?%20_r=2&ref=environment (accessed October 26, 2011).

30 Feldman, S. (2010) Algal fuel inches toward price parity with oil, Reuters, November 22. Available on the Web at: http://www.reuters.com/article/2010/11/22/idUS10859941820101122 (accessed November 20, 2011).

31 Zhang, Q., Ma, J., Qiu, G., et al. (2011) Potential energy production from algae on marginal land in China. Bioresource Technology doi: 10.1016/j.biortech.2011.08.084.

32 Carriquiry, M. A., Du, X., and Timilsina, G. R. (2011) Second-generation biofuels – Economics and policies, World Bank Policy Working Paper 5406, August. Available on the Web at: http://www-wds.worldbank.org/external/default/WDSContentServer/WDSP/IB/2010/08/30/000158349_20100830090558/Rendered/PDF/WPS5406.pdf (accessed November 20, 2011).

33 Doughman, S. D. and Krupanidhi, S. (2008) Is algae oil fuel or nutrition? Everyman's Science 153, 164-168.

34 Rapier, R. (2008) Neste moves forward with green diesel, R-Squared Energy Blog, June 13. Available on the Web at: http://i-r-squared.blogspot.com/2008/06/neste-moves-forward-with-green-diesel.html (accessed October 26, 2011).

35 Wright, M. and Brown, R. C. (2007) Establishing the optimal sizes of different kinds of biorefineries, Biofuels, Bioprocessing, and Biorefineries 1, 191–200.

36 Chynoweth, D., Owens, J. and Legrand, R. (2001) Renewable methane from anaerobic digestion of biomass, Renewable Energy 22, 1-8.

37 Brown, R. C. (2003) Biorenewable Resources: Engineering New Products from Agriculture, Blackwell Publishing, Ames, pp. 152-156.

38 Persson, M., Jonsson, O., and Wellinger, A., (2006) Biogas upgrading to vehicle fuel standards and grid injection, IEA Bioenergy Task 37 Report. Available on the Web at http://www.iea-biogas.net/_download/ publi-task37/upgrading_report_final.pdf (accessed October 26, 2011).

39 Golueke, C. G., Oswald, W. J., and Gotaas, H. B. (1957) Anaerobic digestion of algae, Appl Microbiol. 5, 47–55.

40 Sampson, R. and LeDuy, A. (1983) Improved performance of anaerobic digestion of Spirulina maxima algal biomass by addition of carbon-rich wastes, Biotechnology Letters 5, 677-682.

41 Golueke, C. G. and Oswald, W. J. (1959) Biological Conversion of Light Energy to the Chemical Energy of Methane, Appl Microbiol. 7, 219–227.

Chapter 9

1 Brown, R. C. (2003) Biorenewable Resources: Engineering New Products from Agriculture, Blackwell Publishing, Ames, IA, pp. 159-165.

2 Sjostrom, E. (1993) Wood chemistry: Fundamentals and applications, Second Edition, Academic Press, San Diego, CA.

3 Brown, R. C. (2003) Biorenewable Resources: Engineering New Products from Agriculture, Blackwell Publishing, Ames, IA, pp. 59-61.

4 Graham, R. L., Nelson, R., Sheehan, J., Perlack, R. D., Wright, L. L. (2007) Current and potential U.S. corn stover supplies, Agronomy Journal 99, 1-11.

5 Wilcke, W. and Wyatt, G. (2007) Grain storage tips: Factors and formulas for crop drying, storage and handling, University of Minnesota Extension Service. Available on the web at: http:// www. extension.umn.edu/specializations/cropsystems/M1080-FS. pdf (accessed October 26, 2011).

6 USDA (2011) Feed Grains Database: Yearbook Tables, Available on the web at: http://www.ers.usda.gov/Data/FeedGrains/ (accessed October 26, 2011).

7 Wright, L. L, Hohenstein, Eds. (1994) Dedicated Feedstock Supply Systems: Their Current Status in the USA, Biomass and Bioenergy 6.

8 Ranney, J. W. and Mann, L. K. (1994) Environmental considerations in energy crop production, Biomass and Bioenergy 6(3) 211.

9 Hesser, L. (2006) The Man who Fed the World, Durban House Publishing, Dallas, TX.

10 Brown, R. C. (2003) Biorenewable Resources: Engineering New Products from Agriculture, Blackwell Publishing, Ames, IA, pp. 102-106.

11 McLaughlin, S. B., Kzos, L. A. (2005) Development of switchgrass (Panicum virgatum) as a bioenergy feedstock in the United States, Biomass and Bioenergy 28, 515-535.

12 Sanderson, M. A. and Adler, P. R. (2008) Perennial forages as second generation bioenergy crops, Int J Mol Sci. 9(5): 768–788.

13 Anon (2005) Impacts of modeled recommendations of the National Commission on Energy Policy, Energy Information Administration, April. Available on the web at: http://www.eia.gov/ oiaf/ servicerpt/bingaman/background.html (accessed October 26, 2011).

14 Samson, R., et al. (2008) Developing Energy Crops for Thermal Applications: Optimizing Fuel Quality, Energy Security and GHG Mitigation, Biofuels, Solar and Wind as Renewable Energy Systems: Benefits and Risks, Springer Science, Berlin.

15 Anon (2009) Questions and answers about miscanthus, Bioenergy Feedstock Information Network. Available on the web at: http://bioenergy.ornl.gov/papers/miscanthus/miscanthus.html (accessed October 26, 2011).

16 Möller, Ralf, et al. (2007) Crop Platforms for Cell Wall Biorefining: Lignocellulose Feedstocks, EPOBIO Project, April. Available on the web at: http://www.epobio.net/pdfs/0704LignocelluloseFeedstocks Report.pdf (accessed October 26, 2011).

17 Perlack, R. D., Wright, L. L., Turhollow, A. F., Graham, R. L., Stokes, B. J., and Erbach, D. C. (2005) Biomass as Feedstock for a Bioenergy and Bioproducts Industry: The Technical Feasibility of a Billion-Ton Annual Supply, Department of Energy, DOE/GO-102995-2135, April. Available on the Web at: http://www1.eere.energy.gov/biomass/pdfs/final_billionton_vision_report2.pdf (accessed October 26, 2011).

18 U.S. Department of Energy (2011) U.S. Billion-Ton Update: Biomass Supply for a Bioenergy and Bioproducts Industry, Perlack, R. D. and Stokes, B. J. (leads), ORNL/TM-2011/224, August. Available on the Web at: https://bioenergykdf.net/content/billiontonupdate (accessed November 20, 2011).

19 U.S. Department of Agriculture (2009) Summary Report: 2007 National Resources Inventory, Natural Resources Conservation Service, Washington, DC, and Center for Survey Statistics and Methodology, Iowa State University, Ames, IA. Available on the Web at: http://www.nrcs.usda.gov/Internet/ FSE_DOCUMENTS//stelprdb1041379.pdf (accessed October 26, 2011).

20 Alig, R., et al. (2003) Land use changes involving forestry in the United States: 1952 to 1997, with projections to 2050, TR/ PNW-GTR-587, U.S. Department of Agriculture, Forest Service, Pacific Northwest Research Station, Corvallis, OR, September. Available on the Web at: http://www.fs.fed. us/pnw/pubs/gtr587. pdf (accessed October 26, 2011).

21 Rowell, R., Young, R. and Rowell, J. (1997) Paper and Composites from Agro-Based Resources, CRC Press, Boca Raton.

22 Morrison, I. M. (1979) Carbohydrate chemistry and rumen digestion, Proc. Nutr. Soc. 38, 269-274.

23 Brown, R. C. (2003) Biorenewable Resources: Engineering new products from agriculture, Iowa State Press, Ames, p. 169-179.

24 Lynd, L. R. (1996) Overview and evaluation of fuel ethanol from cellulosic biomass: technology, economics, the environment, and policy, Annu. Rev. Energy Environ. 21, 403–465.

25 Zhang Y-H. P, Lynd L. R. (2004) Toward an aggregated understanding of enzymatic hydrolysis of cellulose: Non-complexed cellulose systems, Biotechnol. Bioeng. 88, 797–824.

26 Dien, B. S., Li, X.-L., Iten, L. B., Jordan, D. B., Nichols, N. N., O'Bryan, P. J., Cotta, M. A. (2006) Enzyme and Microbial Technology 39, 1137–1144.

27 Kirk, O., Borchert T. V., Fuglsang, C. C. (2002) Industrial enzyme applications, Curr. Opin. Biotechnol. 13, 345–351.

28 Moreira, N. (2005) Growing expectations: New technology could turn fuel into a bumper crop, Sci. News Online 168, 209–24.

29 Zhang, Y. H. P., Himmel, M. E., Mielenz, J. R. (2006) Outlook for cellulase improvement: Screening and selection strategies, Biotechnology Advances 24, 452–481.

30 Bai, F. W., Anderson, W.A., Moo-Young, M. (2008) Ethanol fermentation technologies from sugar and starch feedstocks, Biotechnology Advances 26, 89-105.

31 du Preez, J. C., Bosch M., Prior, B. A. (1987) Temperature profiles of growth and ethanol tolerance of the xylose-fermenting yeasts Candida shehatae and Pichia stipitis, Appl. Microbiol. Biotechnol. 25, 521–525.

32 Ingram, L. O., Gomez, P. F., Lai, X., Moniruzzaman, M., Wood, B. E., Yomano, L. P., and York, S. W. (1998) Metabolic engineering of bacteria for ethanol production. Biotechnol. Bioeng. 58, 204–214.

33 Krishnan, M. S., Xia, Y., Ho, N. W. Y., Tsao, G. T. (1997) Fuel ethanol production from lignocellulosic sugars: Studies using a genetically engineered Saccharomyces yeast, ACS Symposium Series 666, 74-92.

34 Ho N. W. Y and Tsao, G. T. (1993) Recombinant yeasts for effective fermentation of glucose and xylose, US Patent No. 08/148, 581.

35 Lynd, L. R. (2004) Consolidated bioprocessing for cellulosic biomass, Industrial Bioprocessing 26, 4.

36 Shapouri, H., Salassi, M., Fairbanks, J. N. (2006) The economic feasibility of ethanol production from sugar in the United States, U.S. Department of Agriculture. Available on the Web at: http:// www. usda.gov/oce/reports/energy/EthanolSugarFeasibilityReport3.pdf (accessed October 26, 2011).

37 DOE (2009) Theoretical ethanol yield calculator, Office of Energy Efficiency and Renewable Energy, Available on the web: http://www1.eere.energy.gov/biomass/ethanol_yield_calculator.html (accessed October 26, 2011).

38 Lynd, L. R. (1996) Overview and evaluation of fuel ethanol from cellulosic biomass: Technology, economics, the environment and policy, Annual Reviews in Energy and the Environment 21, 403-465.

39 Chase, R. (2006) DuPont, BP join to make butanol; they say it outperforms ethanol as a fuel additive, USA Today, June 23. Available on the Web at: http://www.usatoday.com/money/industries/ energy/2006-06-20-butanol_x.htm (accessed October 26, 2011).

40 Wang, B., et al. (2008) Production of acetone-butanol-ethanol (ABE) with distiller's dried grains with solubles, American Society of Agricultural and Biological Engineers, St. Joseph.

41 Ladygina, N., Dedyukhina, E., and Vainshtein, M. (2006) A review on microbial synthesis of hydrocarbons, Process Biochemistry 41, 1001-1014.

42 Baraud, J., et al. (1967) Présence d'hydrocarbures dans l'insaponifiable des lipides de levures, CR Acad. Sci. (Paris) Ser. D 265, 83-85.

43 Jones, J. (1969) Studies on lipids of soil micro-organisms with particular reference to hydrocarbons. J. Gen. Microbiol. 59, 145-52.

44 Wu, S., et al. (2006) Redirection of cytosolic or plastidic isoprenoid precursors elevates terpene production in plants, Nature Biotechnology 24, 1441-1447.

45 Brown, R.C. (2003) Biorenewable Resources: Engineering new products from agriculture, Iowa State Press, Ames, p. 30-32.

Chapter 10

1 NSF (2008) Breaking the Chemical and Engineering Barriers to Lignocellulosic Biofuels: Next Generation Hydrocarbon Biorefineries, Ed. Huber, G., U.S Department of Energy. Available on the Web at: http://www.ecs.umass.edu/biofuels/Images/Roadmap2-08.pdf (accessed October 26, 2011).

2 Lammers, D. (2007) Gasification may be key to U.S. ethanol, Associated Press, March 4. Available on the Web at: http://www.physorg.com/news92295273.html (accessed October 26, 2011).

3 Regalbuto, J. (2011) The sea change in US biofuels' funding: from cellulosic ethanol to green gasoline, Biofuels, Bioproducts and Biorefining 5, 495-504.

4 Brown, R. C. (2003) Biorenewable Resources: Engineering new products from agriculture, Iowa State Press, Ames, pp. 145-146.

5 Higman, C. and Van Der Burgt, M. (2008) Gasification, Gulf Professional Publishing, Oxford, pp. 1-3.

6 Henrich, E. and Weirich, F. (2004) Pressurized entrained flow gasifiers for biomass, Environmental Engineering Science 21, 53-64.

7 Dayton, D. C., Turk, B., and Gupta, R., (2011) Syngas cleanup, conditioning, and utilization, in Thermochemical Processing of Biomass, R. C. Brown, Ed., Wiley, Chichester, U.K., pp. 78-123.

8 Bridgwater, A. (1995) The technical and economic feasibility of biomass gasification for power generation, Fuel 74, 631-653.

9 Milne, T., Evans, R., and Abatzaglou, N. (1998) Biomass gasifier "tars": Their nature, formation, and conversion, NREL/TP-570-25357, National Renewable Energy Laboratory, Golden, CO. Available on the Web at: http://www.osti.gov/bridge/product.biblio.jsp?osti_id=3726 (accessed October 26, 2011).

10 Zhang, R., Brown, R., Suby, A., and Cummer, K. (2004) Catalytic destruction of tar in biomass-derived producer gas, Energy Conversion and Management 45, 995-1014.

11 Myers, D. B., Ariff, G. D., James, B. D., and Kuhn, R. C. (2003) Hydrogen from renewable energy sources: Pathway to 10 Quads for transportation uses in 2030 to 2050, Final Report to the U.S. Department of Energy, Hydrogen Program Office, Washington, D.C. Available on the Web at: http://www.eere.energy.gov/hydrogenandfuelcells/pdfs/iia11_myers.pdf (accessed October 26, 2011).

12 Bowen, D., Lau, F., Dihu, R., Doong, S., Remick, R., Slimane, R., Zabransky, R., Hughes, E., and Turn, S. (2003) Techno-economic analysis of hydrogen production by gasification of biomass, Final Report to the U.S. Department of Energy, Contract DE-FC36-01GO11089, June. Available on the Web at: http://www.nrel.gov/docs/fy11osti/46587.pdf (accessed October 26, 2011).

13 Spath, P. L, Lane, J. M., Mann, M. K., and Amos, W. A. (2001) Update of hydrogen from biomass: Determination of the delivered cost of hydrogen, National Renewable Energy Laboratory, July. Available on the Web at: http://www.nrel.gov/docs/fy04osti/33112.pdf (accessed October 26, 2011).

14 Brown, R. C., Smeenk, J., Sadaka, S., Norton, G., Zhang, R., Suby, A., Cummer, K., Ritzert, J., Xu, M., Lysenko, S., Nunez, J., and Brown, N. (2005) Biomass-derived hydrogen from a thermally ballasted gasifier, U.S. DOE Final Report, Contract No. DE-FC36-01GO11091, August. Available on the Web at: http://www1.eere.energy.gov/biomass/pdfs/thermally_ballasted_gasifier.pdf (accessed October 26, 2011).

15 Wright, M. and Brown, R. C. (2007) Comparative economics of biorefineries based on the biochemical and thermochemical platforms, Biofuels, Bioprocessing, and Biorefineries 1, 49-56.

16 National Research Council (2004) The hydrogen economy: Opportunities, Costs, Barriers, and R&D Needs, The National Academies Press, Washington, D.C., February, pp. 99-105. Available on the Web at: http://www.nap.edu/openbook.php?isbn=0309091632 (accessed October 26, 2011).

17 Zhang, R., Cummer, K., Suby, A., and Brown, R. C. (2005) Biomass-derived hydrogen from an air-blown gasifier," Fuel Processing Technology 86, 861-874.

18 MacKenzie, J. and Avery, W. (1996) Ammonia fuel: The key to hydrogen-based transportation, Energy Conversion Engineering Conference, Proceedings of the 31st Intersociety 3, 1761-1766.

19 Satterfield, C. (1991) Heterogeneous Catalysis in Industrial Practice, Krieger Publishing Co., Malabar.

20 Rafiqul, I., Weber, C., Lehmann, B., and Voss, A. (2005) Energy efficiency improvements in ammonia production - perspectives and uncertainties, Energy 30, 2487-2504.

21 Higman, C. and van der Burgt, M. (2003) Gasification, Elsevier Science, Amersterdam, pp. 258-259.

22 Probstein, R. F. and Hicks, R. E. (2006) Synthetic Fuels, Dover Publications, Inc., Mineola, NY, pp. 191-194.

23 Davenport, B. (2002) Methanol, in Chemical Economics Handbook Marketing Research Report, SRI International, Menlo Park, CA.

24 Semelsberger, T., Borup, R. L., and Greene, H. (2006) Dimethyl ether (DME) as an alternative fuel, Journal of Power Sources 156, 497-511.

25 Klass, D. (1998). Biomass for Renewable Energy, Fuels, and Chemicals, Academic Press, San Diego: 427-429.

26 Parker, M. (2011) Range fuels cellulosic ethanol plant fails, U.S. Pulls Plug, Bloomberg, December 2. Available on the Web at: http://www.bloomberg.com/news/2011-12-02/range-fuels-cellulosic-ethanol-plant-fails-as-u-s-pulls-plug.html (accessed December 4, 2011).

27 Anon (2005) Fischer-Tropsch archive. Available on the Web at: www.fischer-tropsch.org (accessed April 10, 2009).

28 Swanson, R., Platon, A., Satrio, J., Brown, R.C. (2010) Technoeconomic analysis of biomass-to-liquids production based on gasification, Fuel 89, Supplement 1, S11-S19.

29 Brown, R. C. (2007) Hybrid thermochemical/biological processing, Applied Biochemistry and Biotechnology 137, 947-956.

30 Grethlein, A. and Jain, M. (1992) Bioprocessing of coal-derived synthesis gases by anaerobic bacteria, Trends in Biotechnology 10, 418-423.

31 Anon (2008) Lowest Cost of Production, Coskata Inc. Available on the Web at: http://www.coskata. com/process/index.asp?source=3A0B7092-93E7-452F-9A2A-B87A3BDECA2F (accessed October 26, 2011).

32 Venderbosch, R. H. and Prins, W. (2011) Fast pyrolysis, in Thermochemical Processing of Biomass, R. C. Brown, Ed., Wiley, Chichester, U.K., pp. 124-156.

33 Bridgwater, A. (2003) Renewable fuels and chemicals by thermal processing of biomass, Chemical Engineering Journal 91, 87-102.

34 Brown, R. C. (2003) Biorenewable Resources: Engineering new products from agriculture, Iowa State Press, Ames, pp. 182-186.

35 Oasmaa, A., Peacocke, C., and Tutkimuskeskus, V. T. (2001) A Guide to Physical Property Characterisation of Biomass-derived Fast Pyrolysis Liquids, Technical Research Centre of Finland. Available on the Web at: http://www.vtt.fi/inf/pdf/publications/2001/P450.pdf (accessed October 26, 2011).

36 Patwardhan, P. R., Satrio, J.A., Brown, R. C. and Shanks, B. H. (2009) Product distribution from fast pyrolysis of glucose-based carbohydrates, Journal of Analytical and Applied Pyrolysis 86, 323-330.

37 Patwardhan, P., Satrio, J., Brown, R. and Shanks, B. (2010) Influence of inorganic salts on the primary pyrolysis products of cellulose, Bioresource Technology 101, 4646-4655.

38 Ouellette, R. J. (1994) Organic Chemistry: A Brief Introduction, 2nd Edition, Prentice Hall, NJ, p. 239

39 Piskorz, J., Majerski, P., Radlein, D. (1999) in Biomass—A Growth Opportunity in Green Energy and Value-added Products, Overend, R. P. and Chornet, E. (Eds.), Elsevier Science, Amsterdam, p. 1153.

40 Dalluge, D., Brown, R. C. (2011) Pyrolytic Pathways to Increasing Lignin-Derived Monomer/Oligomer Ratio in Bio-oil, International Conference on Thermochemical Biomass Conversion Science, Chicago, IL, September 27-30.

41 Bayerbach, R. and Meier, D. (2009) Characterization of the water-insoluble fraction from fast pyrolysis liquids (pyrolytic lignin). Part IV: Structure elucidation of oligomeric molecules, J. Anal. Appl. Pyrolysis 85, 98–107.

42 Adler, E. (1977) Lignin chemistry – past, present and future, Wood Science Technology 11, 169-218.

43 Huber, G. W., Chheda, J. N., Barrett, C. J., and Dumesic, J. A. (2005) Production of Liquid Alkanes by Aqueous-Phase Processing of Biomass-Derived Carbohydrates, Science 308, 1446-1450.

44 Holmgren, J., Nair, P., Elliot, D., Bain, R., and Marinangeli, R. (2008) Converting Pyrolysis Oils to Renewable Transport Fuels: Processing Challenges & Opportunities, National Petrochemical & Refiners Association Annual Meeting, San Diego, CA.

45 Wright M., Daugaard D. E., Satrio J. A., and Brown R. C. (2010) Techno-economic analysis of biomass fast pyrolysis to transportation fuels, Fuel 89, S2-S10.

46 Carlson, T., Vispute, T., and Huber, G. (2008) Green Gasoline by Catalytic Fast Pyrolysis of Solid Biomass Derived Compounds, Chem. Sus. Chem 1, 397-400.

47 French, R. and Czernik, S. (2010) Catalytic pyrolysis of biomass for biofuels production, Fuel Processing Technology 91, 25-32.

48 Brown, R. C., Radlein, D., and Piskorz, J. (2001) Pretreatment processes to increase pyrolytic yield of levoglucosan from herbaceous feedstocks, Chemicals and Materials from Renewable Resources, ACS Symposium Series 784, American Chemical Society, Washington, D.C., 2001, pp. 123 - 132.

49 Brown, R. C. (2011) Prospects for a thermolytic sugars platform, International Conference on Thermochemical Biomass Conversion Science, Chicago, IL, September 27-30.

50 So, K. S. and Brown, R. C. (1999) Economic analysis of selected lignocellulose-to-ethanol conversion technologies, Applied Biochemistry and Biotechnology 77, 633-640.

51 Lewkowski, J. (2001) Synthesis, chemistry and applications of 5-hydroxymethylfurfural and its derivatives. ARKIVOC 1, 17-54.

52 Hanniff, M. I. and Dao, L. H. (1987) Conversion of biomass carbohydrates into hydrocarbon products, Inst of Gas Technology, Chicago.

53 Roman-Leshkov, Y., Barrett, C. J., Liu, Z. Y., and Dumesic, J. A. (2007) Production of dimethylfuran for liquid fuels from biomass-derived carbohydrates, Nature 447, 982-986.

54 Zhao, H., et al. (2007) Metal Chlorides in Ionic Liquid Solvents Convert Sugars to 5-Hydroxymethyl-furfural, Science 316, 1597.

55 Kunkes, E. L., Simonetti, D. A., West, R. M., Serrano-Ruiz, J. C., Gartner, C. A., and Dumesic, J. A. (2008) Catalytic Conversion of Biomass to Monofunctional Hydrocarbons and Targeted Liquid-Fuel Classes, Science 322, 417-421.

56 Elliott, D., Hydrothermal processing, in Thermochemical Processing of Biomass, R. C. Brown, Ed., Wiley, Chichester, U.K., pp. 200-231.

57 Elliott, D., et al. (1991) Developments in direct thermochemical liquefaction of biomass: 1983-1990. Energy & Fuels 5, 399-410.

58 Elliott, D., et al. (2004) Chemical processing in high-pressure aqueous environments: Process development for catalytic gasification of wet Biomass feedstocks, Industrial and Engineering Chemistry Research 43, 1999-2004.

59 Anon (2009) Solarbuzz, Photovoltaic Industry Statistics, Available on the Web at: http://www.solarbuzz.com/facts-and-figures/markets-growth/cost-competitiveness (accessed October 26, 2011).

60 Jenkins, B. M, Baxter, L. L, and Koppejan, J. (2011) Biomass combustion, in Thermochemical Processing of Biomass, R. C. Brown, Ed., Wiley, Chichester, U.K., pp. 13-46.

61 Brown, R.C. (2003) Biorenewable Resources: Engineering new products from agriculture, Iowa State Press, Ames, pp. 68-69.

62 Johansson, T. B., Kelly, H., Reddy, A.K.N., and Williams, R. H. (1993) Renewable fuels and electricity for a growing world economy, pp. 172, in Renewable Energy: Sources for Fuels and Electricity.

63 Williams, R. H. and Larson, E. D. (1996) Biomass gasifier gas turbine power generating technology, Biomass and Bioenergy 10, 149-166.

64 Poullikkas, A. (2005) An overview of current and future sustainable gas turbine technologies, Renewable and Sustainable Energy Reviews 9(5): 409-443.

65 Dicks, A. and Larminie, J. (2000) Fuel Cell Systems Explained, John Wiley & Sons.=

Chapter 11

1 Charles, D. (2009) Stimulus Gives DOE Billions for Carbon-Capture Project, Science 323, 1158.

2 Klemesa, J., Bulatova, I., and Cockerill, T. (2007) Techno-economic modeling and cost functions of CO2 capture processes, Computers & Chemical Engineering 31, 445-455.

3 Hamilton, M. R. Herzog, Howard J.; Parsons, J. E. (2009) Cost and U.S. public policy for new coal power plants with carbon capture and sequestration, Energy Procedia 1, 4487-4494.

4 Anon (2010) This changes everything, The Economist, March 13-19, pp.16-18.

5 U.S. DOE (2010) Fuels from Sunlight, Available on the Web at: http://energy.gov/articles/fuels-sunlight-hub (accessed November 23, 2011).

6 Anon (2009) Five More Airlines Join Sustainable Aviation Fuel Users Group, Green Car Congress, July 13. Available on the Web at: http://www.greencarcongress.com/2009/07/safug-20090713.html (accessed November 23, 2011).

7 Bergin, T. and Driver, A. (2009) Exxon to try to develop biofuel from algae, Reuters, Jul 14. Available on the Web at: http://www.reuters.com/article/GCA-BusinessofGreen/idUSTRE-56D0O120090714 (accessed November 23, 2011).

8 Sheehan, J., Dunahay, T., Benemann, J., Roessler, P. (1998) A look back at the U.S. Department of Energy's aquatic species program—Biodiesel from algae, National Renewable Energy Laboratory Technical Report, NREL/TP-580-24190.

9 For example: Burnham, M. (2009) DOE Earmarks $85M for 'Advanced' Biofuels, July. Available on the Web at: http://www.nytimes.com/gwire/2009/07/17/17greenwire-doe-earmarks-85m-for-advanced-biofuels-78090.html (accessed November 23, 2011).

10 Anon (2008) Air NZ Test Flight Proves Viability of Biofuel, Aviation Industry Environmental News, December. Available on the Web at: http://www.enviro.aero/Aviationindustryenvironmentalnews. aspx?NID=306 (accessed November 23, 2011).

11 EIA (2011) U.S. fuel ethanol plant production capacity, Energy Information Agency, November 29. Available on the Web at: http://www.eia.gov/petroleum/ethanolcapacity/ (accessed December 6, 2011).

12 Anon (2009) California Air Resources Board votes to recognize sugar cane ethanol's carbon reduction levels, World Wire, April 23. Available on the Web at: http://world-wire.com/news/0904230003.html (accessed November 23, 2011).

13 Keeney, D.R. and T.H. DeLuca (1992) Biomass as an energy source for the Midwestern, U.S. Am. J. Alternative Agric. 7, 137-144.

14 Wright, M. M., Brown, R. C., Boateng, A. A. (2008) Distributed processing of biomass to bio-oil for subsequent production of Fischer-Tropsch liquids, Biofuels, Bioprocessing, and Biorefineries 2, 229-238.

15 Marker, T. L., et al. (2005) Opportunities for biorenewables in oil refineries, U.S. Department of Energy Final Report, Prepared by UOP, Inc., December 12.

16 Anon (2006) ConocoPhillips Begins Production of Renewable Diesel Fuel at Whitegate Refinery, Green Car Congress, December 20. Available on the Web at: http://www.greencarcongress.com/ 2006/12/conocophillips_.html (accessed November 23, 2011).

17 Lane, J. (2010) Jet Stream: Biofuels Digest Special report on Aviation Biofuels, Biofuels Digest, March 3. Available on the Web at: http://www.biofuelsdigest.com/bdigest/2010/03/03/jet-stream-biofuels-digest-special-report-on-aviation-biofuels/ (accessed November 23, 2011).

18 H.R. 6 Energy Independence and Security Act of 2007; available on the Web at http://www.govtrack.us/ congress/billtext.xpd?bill=h110-6 (accessed November 23, 2011).

19 Kline, K. L. and Dale, V. H. (2008) Biofuels: Effects on land and fire; Letter to the editor, Science 321, 199.

20 Wiedmann, T., Wood, R., Lenzen, M., Minx, J., Guan, D. and Barrett, J. (2008) Development of an Embedded Carbon Emissions Indicator - Producing a Time Series of Input-Output Tables and Embedded Carbon Dioxide Emissions for the UK by Using a MRIO Data Optimisation System, Final Report to the Department for Environment, Food and Rural Affairs, Prepared by the Stockholm Environment Institute at the University of York and the Centre for Integrated Sustainability Analysis at the University of Sydney.

21 Brown, T., Hayes, D., and Brown, R. (2009) The Embedded Carbon Valuation System: A Policy Concept to Address Climate Change, Farm Foundation. Available on the Web at: http://www.farmfoundation.org/news/articlefiles/1718-Brown%20Hayes%20and%20Brown.pdf (accessed November 23, 2011).

22 EIA (2006) World Carbon Intensity--World Carbon Dioxide Emissions from the Consumption and Flaring of Fossil Fuels per Thousand Dollars of Gross Domestic Product Using Market Exchange Rates, 1980-2006International Energy Annual, U.S. Department of Energy, Energy Information Agency, available on the Web at: http://www.eia.doe.gov/environment.html (accessed November 23, 2011).

23 Peters, G. P., Minx, J. C., Weber, C. L., and Edenhofer, O. (2011) Growth in emission transfers via international trade from 1990 to 2008, Proceedings of the National Academy of Sciences 108, 8903-8908.

24 Anon (2008) Emissions of Greenhouse Gases Report, Energy Information Administration, DOE/EIA-0573(2007), December 3. Available on the web at: http://www.eia.gov/oiaf/1605/ggrpt/carbon. html (accessed November 23, 2011).

25 Anon (2009) The Alternative Choice, The Economist, July 4, p. 64.

26 Albuisson,M., Lefèvre, M., Wald, L. (2006) Yearly Mean of Irradiance in the World, Centre for Energy and Processes, Ecole des Mines de Paris. Available on the Web at: http://www.soda-is.com/img/map_ed_13_world.pdf (accessed November 23, 2011).

27 FAO, Annual Average Rainfall Total, Climate Impact on Agriculture. Available on the Web at: http://www.fao.org/nr/climpag/climate/img/2rainavg.gif (accessed November 23, 2011).

28 FAO, Potential Biomass, Climate Impact on Agriculture. Available on the Web at: http://www.fao.org/nr/climpag/climate/img/3biomass.gif (accessed November 23, 2011).

29 United Nations, Human Development Reports, United Nations Development Program. Available on the Web at: http://hdr.undp.org/en/ (accessed November 23, 2011).

30 USDA (2005) Global Soil Regions Map, Natural Resources Conservation Services. Available on the Web at: http://soils.usda.gov/use/worldsoils/mapindex/order.html (accessed November 23, 2011).

31 Marris, E., Black is the new green, Nature 442, 624-626, August 10, 2006.

32 Woods, W .I. and McCann, J. M. (1999) in Yearbook Conf. Latin Am. Geogr. 25, 7–14, Caviedes, C., ed. Univ. Texas, Austin.

33 Zech, W., Haumaier, L., and Hempfling, R. (1990) Ecological aspects of soil organic matter in tropical land use, in Humic Substances in Soil and Crop Sciences: Selected Readings, ed. by McCarthy, P., Clapp, C. E., Malcolm, R. L. and Bloom, P. R. Soil Sci Soc Am, Madison, WI, pp. 187–202.

34 Liang, B., Lehmann, J., Solomon, D., Kinyangi, J., Grossman, J., O'Neill, B., Skjemstad, J. O., Thies, J., Luiza, F. J., Petersen, J., and Neves, E. G. (2006) Black carbon increases cation exchange capacity in soils. Soil Sci Soc Am J 70, 1719–1730.

35 Warnock, D. D, Lehmann, J., Kuyper, T. W., and Rillig, M. C. (2007) Mycorrhizal responses to biochar in soil – concepts and mechanisms, Plant Soil 300, 9–20.

36 Glaser, B., Haumaier, L., Guggenberger, G., and Zech, W. (2001) The 'Terra Preta' phenomenon: a model for sustainable agriculture in the humid tropics, Naturwissenschaften 88, 37–41.

37 Lehmann, J. Kaampf, N. Woods, W.I. Sombroek, W. Kern, D.C. Cunha T.J.F. (2003) Historical Ecology and Future Explorations in Amazonian Dark Earths: origin, properties, and management, Chapter 23, p. 484.

38 D.A. Laird, Brown, R.C., Amonette, J.E., Lehmann J. (2009) Review of the pyrolysis platform for producing bio-oil and biochar: Technology, logistics and potential impacts on greenhouse gas emissions, water quality, soil quality and agricultural productivity, Biofuels, Bioproducts, and Biorefining 3, 547–562.

39 Brown R. (2009) Biochar production technology, in Biochar for Environmental Management, ed. by Lehmann, J. and Joseph, S., Earthsan, London, pp. 127-146.

40 Namaalwa, J., Sankhayan, P. L., and Hofstad, O. (2007) A dynamic bio-economic model for analyzing deforestation and degradation: An application to woodlands in Uganda, Forest Policy Econ 9, 479-495.

41 Anex, R. P., Lynd, L. R., Laser, M. S., Heggenstaller, A. H., and Liebman, M. (2007) Potential for enhanced nutrient cycling through coupling of agricultural and bioenergy systems. Crop Sci 47, 1327–1335.

42 United Nations Environment Programme (UNEP), Global Environment Outlook (Oxford University Press, New York, 1997), p. 236.

Index

Made in the USA
Lexington, KY
14 August 2012